African Mathematics

From Bones to Computers

Abdul Karim Bangura

UNIVERSITY PRESS OF AMERICA,® INC.
Lanham • Boulder • New York • Toronto • Plymouth, UK

University Press of America,® Inc.
4501 Forbes Boulevard
Suite 200
Lanham, Maryland 20706
UPA Acquisitions Department (301) 459-3366

Estover Road
Plymouth PL6 7PY
United Kingdom

British Library Cataloging in Publication Information Available

Library of Congress Control Number: 2010934250
ISBN: 978-0-7618-5348-0 (paperback : alk. paper)
eISBN: 978-0-7618-5349-7

To the Afrikan teacher and learner!

Contents

Chapter One

General Introduction

Scholars and other professionals working in the field of Mathematics Education in Africa have identified a plethora of problematic issues in the endeavor. These issues include attitudes, curriculum development, educational change, instruction, academic achievement, standardized and other tests, performance factors, student characteristics, cross-cultural differences and studies, literacy, native speakers, social class and differences, equal education, teaching methods, knowledge level, educational guidelines and policies, teacher associations, transitional schools, comparative education, other subjects such as Physics and Social Studies, skills development, surveys, talent, educational research, teacher education and qualifications, academic standards, teacher effectiveness, lesson plans and modules, teacher relationship, teacher characteristics, instructional materials, program effectiveness, program evaluation, African culture, African history, Black Studies, class activities, educational games, number systems, cognitive ability, foreign influence, inequalities, ethnicities, and fundamental concepts (Adler, 1994; African-American Institute, 1976; Ginsburg, 1978; Hoadley, 2007; Howie, 1997; Howie and Hughes, 1998; Howie and Pietersen, 2001; Howie et al., 2000; Linder and Hudson, 1989; Jama, 1983; Le Roux et al., 1985; Masota, 1982; Mbiriru, Sallah, 1982; 1983; Williams, 1978; Zaaiman et al., 2000; Zaslavsky, 1970). While this book is not intended to serve as a panacea for these problems, for that requires a different work, it is our hope, however, that it will be of use to professionals in Mathematics Education in addressing at least a few of these issues.

In its extensive 2007 study titled *Developing Science, Mathematics, and ICT Education in Sub-Saharan Africa: Patterns and Promising Practices*, authored by Wout Ottevanger, Jan van den Akker and Leo de Feiter, the World Bank in its effort to assist African countries south of the Sahara to overcome the difficulties encountered in Science, Mathematics, and Information and

Communication Technology (SMICT) education in secondary schools offers a number of suggestions. The study is based on ten Sub-Saharan African nations: Botswana, Burkina Faso, Ghana, Namibia, Nigeria, Senegal, South Africa, Uganda, Tanzania, and Zimbabwe. The major challenges that the authors consider to impede SMICT education in those countries are poorly-resourced schools, large classes, a curriculum that is hardly relevant to students' daily lives, a lack of qualified teachers, and inadequate teacher education programs (2007:v).

The World Bank study goes on to say this about learning Science and Mathematics in Africa south of the Sahara:

> Science and mathematics have always been considered difficult subjects. Many science and mathematics concepts are counter-intuitive and therefore difficult to learn; in fact, often students do not succeed and are then stuck with misconceptions . . . As science and mathematics learning and the curriculum spiral in the sense that at higher educational levels content is treated at higher levels of formalization and abstraction, SMICT education is built on shaky foundations and students experience problems in applying their knowledge in practical contexts. Paradigmatic in this sense are well-known examples of university students who still have major problems with basic mathematical operations like fractions and decimal points, although they may have—more or less successfully—gone further with much more advanced topics (2007:40).

Despite this clearly stated problem by the World Bank study, nowhere in the study is it mentioned that either African Mathematics or African Science can be part and parcel of the remedy.

This book attempts to provide a comprehensive examination of African Mathematics, by tracing the subject from its early beginnings with bones on to the contemporary computer era. Perhaps the following excerpt from the synopsis of Ifeoma Onyefulu's book, *A Triangle for Adaora: An African Book of Shapes* (2007), captures at least one aspect of the essence of African Mathematics in everyday life:

> In the center of Adaora's slice of *paw-paw* is a perfect star shape. She doesn't want to spoil it, so she and her cousin Ugo set off to find a different piece of fruit. As they walk, the children see all kinds of shapes: Uncle Eze's rectangular *agbada*, musicians playing circle-topped elephant drums, a crescent-shaped plantain, even plants with leaves in the shape of a heart.

That Mathematics—generally defined as the systematic study of quantities and relations through the use of numbers and symbols—pervades every branch of human knowledge is hardly a matter of dispute. It is a useful and fascinating field of inquiry, and it possesses the power to solve some of the deepest puzzles humans encounter.

Mathematics is used in everyday life in such simple ways as telling time or counting the change returned by a cashier. A customer in a supermarket employs Mathematics when s/he buys groceries. A wife and husband use Mathematics to make their household budget or to keep track of their bank accounts. And children use Mathematics in many of the games they play.

In business, Mathematics is prevalent in all exchange transactions. Business entities need Mathematics to establish and maintain their records. Bankers rely on Mathematics to handle and invest money. Many businesses rely upon accountants to keep their records and statisticians to analyze large aggregates of data. And insurance companies rely upon actuaries to compute the rates charged for insurance.

In industry, almost all companies rely upon Mathematics for their research and development. Many major industrial firms hire trained mathematicians. All engineering projects rely upon Mathematics. For instance, designing a superhighway requires a great deal of mathematical analyses. Without extensive mathematical formulae and calculations, constructing a giant dam would be impossible. The fact that engineering students must take many Mathematics courses highlights the importance of this field of study.

In science, Mathematics pervades all aspects of the field. Without exact mathematical descriptions, formulae and observations, most scientists will not be able to perform their tasks. Many scientific problems have become so complex that without a sound training in Mathematics, one may not be able to solve them. The hard, behavioral and social sciences all depend on Mathematics to advance their disciplines.

Despite this extensive utilization of Mathematics, it is not obvious to many that it is also a tool used to teach students that Europeans are culturally superior to Africans. Most books dealing with the history of Mathematics devote only a few pages to ancient Egypt and to northern Africa during the Middle Ages. These books generally ignore the history of Mathematics in Africa south of the Sahara, giving the impression that this history either did not exist or, at least, is not knowable, traceable, or, even more absurd, that there was no Mathematics at all in that part of Africa. To Eurocentric scholars, even the Africanity of Egyptian Mathematics is often denied. But contrary to the popular Eurocentric view, one can neither racially nor geographically separate Egyptian civilization from its Black African roots.

That Africa was in the center of Mathematics history for tens of thousands of years is undisputable. From the civilizations across the continent emerged contributions which would enrich both ancient and modern understanding of nature through Mathematics. From the measurement used in the African forest kingdoms to the Mathematics used in building the great stone complexes of Zimbabwe, the efficient irrigation technologies, central administration, the great accuracy of the dimensions of the pyramids, and the random number generation of the binary code that led to the invention of the computer, the achievements of Africans remain a fascination.

Since the 1970s, many books and articles on different topics dealing with African Mathematics have been published, albeit only six of the books are extensive enough for review as we did in the following section of this chapter. What we provide in this book are various aspects of Africa's contributions to Mathematics. This, we hope, will give the reader a comprehensive picture of African Mathematics. The reader interested in greater details of a particular topic will be well served by the relevant references.

BOOKS ON AFRICAN MATHEMATICS

As can be seen in the Bibliography of this book, while many works have been done on African Mathematics, only a handful of books have been published. Of these books, we found only six of them to be extensive for review. These books are, however, somewhat limited in that each of them examines only an aspect of African Mathematics. The following is a review of these books in the chronological order in which they were published.

The first extensive work on African Mathematics is Claudia Zaslavsky's book, *Africa Counts: Number and Pattern in African Cultures* (1973/1999). It is an excellent introduction to the subject. Zaslavsky presents an overview of the available literature in the history of African Mathematics south of the Sahara. She also discusses numbers in terms of words, gestures and their significance; numbers in daily life; mathematical recreations; pattern and shape; regional mathematical studies of Southwest Nigeria and East Africa; past and future aspects of pure Mathematics on the continent. The strength of the book hinges upon how Zaslavsky skillfully combines her expertise in the field of Mathematics and African History to provide a scholarly and well documented account of Africa's contribution to the field of Mathematics.

Although ignored by mainstream mathematicians, Deborah Lela Moore's book titled *The African Roots of Mathematics* (1994) is an excellent account of the Egyptian roots of Mathematics. For example, she shows that the doorways of many of the massive temples in Kemet (ancient Egypt) are shaped in the symbol of Pi. According to her, one of the oldest mathematical documents in existence is the Ahmose Mathematical Papyrus called "The Directions For Knowing All Dark Things" that reveals how the Egyptians derived the

formula for Pi. This document is referred to today as the Rhind Mathematical Papyrus.

Every once in a while, a book is published that significantly advances scholarship. Thus, we are not going to be apologetic for stating boldly that Ron Eglash's *African Fractals: Modern Computing and Indigenous Design* (1999) is bound to be regarded as one of the greatest books on African Mathematics produced in this century. The book is divided into three parts with 14 chapters. The first part introduces Fractal Geometry for people without any Mathematics background, fractals in African settlement architecture, fractals in cross-cultural comparison, and intention and invention in design. The second part discusses geometric algorithms, scaling, numeric systems, recursion, infinity, and complexity. The third part focuses on theoretical frameworks in cultural studies of knowledge, the politics of African fractals, fractals in European history of Mathematics, and futures for African fractals.

What Eglash teaches in the 14 chapters is that elaborate cornrow braids on an African woman's head, for example, can be viewed as more than an affinity with culture or a fashion statement. The intricate patterns are also useful for learning about African fractals—geometric patterns that are repeated on smaller and smaller scales to produce intricate designs that are beyond the scope of classical or Euclidean Geometry. Fractal Geometry has emerged as one of the most exciting frontiers in the fusion between Mathematics and Information Technology. Fractals can be observed in many of the swirling patterns produced by computer graphics, and they have become a vital tool for modeling in the natural sciences. While Fractal Geometry can allow one to get into the far reaches of high tech science, its patterns are surprisingly common in traditional African designs. Also, some of the basic concepts in Fractal Geometry are fundamental to African knowledge systems: quantitative techniques, symbolic systems, engineering, architecture, games, traditional hairstyling, textiles, sculpture, painting, carving, metalwork, and religion. As Eglash explains, although most people learn Euclidean Geometry in school, few study Fractal Geometry, which plays a significant role in the computer modeling process in the hard sciences. Meanwhile, according to Eglash, Fractal Geometry has long been a theme in Africa, with a wide variety of local cultural associations, including kinship, labor practices, politics, and religion.

Eglash's research began in the 1980s while investigating settlement architecture in Central and West Africa. Aerial photographs of various settlement compounds revealed that many were composed of circular structures enclosed in other circles, or rectangles within rectangles, and that the compounds were likely to have street patterns in which broad avenues branched into very small footpaths. As Eglash notes, at first, he thought it was just from unconscious social dynamics. But during his fieldwork, he found that fractal designs also appear in a wide variety of intentional designs—carving, hairstyling, metalwork, painting, textiles—and the recursive process of fractal algorithms are even employed in African quantitative systems. Eglash adds that in the design rationales and cultural semantics of many African geometric figures, as well as in indigenous quantitative systems (additive progression, doubling sequences, binary recursion) and symbolic systems (iconic symbols for feedback loops, equiangular spirals, infinity), there are abstract ideas and formal structures that closely parallel some of the fundamental aspects of Fractal Geometry. These results, Eglash concludes, are congruent with recent developments in complex systems theory, which suggest that pre-modern, non-state societies were neither utterly anarchic, nor frozen in static order, but rather utilized an adaptive flexibility that capitalized on the nonlinear aspects of ecological dynamics.

While in Africa, Eglash encountered some of the most complex fractal systems that exist in religious activities, such as the sequence of symbols used in sand divination, a method of fortune telling found in Senegal. Some of his other findings include the use of sophisticated mathematical ideas in everyday objects. In the arid region of the Sahel, for example, artisans produce windscreens by utilizing a scaling design that gives them the maximum effect—keeping out the wind-driven dust—for the minimum amount of effort and material.

When Eglash returned from Africa, one of his colleagues advised him to focus on scaling patterns in African hairstyles. An enthusiastic group of students at Evergreen State University volunteered their programming skills to help create a multimedia lesson on African fractals. The Hairstyle Storyboard Web site that has been developed utilizes a style referred to as "the braids of threads," from Yaoundé, Cameroon, to explicate African branching fractals. The "fractal hairstyle" module guides users, step by step, through the creation of a three-dimensional fractal, beginning with the initial design and then mathematically determining the ratio of each iteration. The major goal is to eventually combine the images, software and video on African fractals.

Given all this, at least two critical questions can be raised: (1) Since some scholars have found that all cities (historic, primitive and modern) are fractal precisely because they are complex natural systems, and other scholars have discovered that fractal tiling patterns exist on some of the oldest European tiled floors and in ancient Chinese art, what then does this say for the validity of Eglash's arguments concerning African fractals? (2) At what number of scales does self-similarity occur in African fractals and what method does Eglash employ to determine self-similarity? Eglash deals with these questions in several ways.

First, Eglash demonstrates that traditional African settlements typically show repetition of similar patterns at ever-diminishing scales: circles of circles of circular dwellings, rectangular walls enclosing ever-smaller rectangles, and streets in which broad avenues branch down to tiny footpaths with striking geometric repetition. He easily identifies the fractal structure when he compares aerial views of African villages and cities with corresponding fractal graphics simulations. To estimate the fractal dimension of a spatial pattern, Eglash uses several different approaches. In the case of Mokoulek, for instance, which is a black-and-white architectural diagram, a two-dimensional version of the ruler size versus length plots is employed. However, for the aerial photo of Labbazanga, an image in shades of gray, the Fourier Transform is used. Nonetheless, according to Eglash, we cannot just assume that African fractals show an understanding of Fractal Geometry, nor can we dismiss that possibility. Thus, he insists that we listen to what the designers and users of these structures have to say about it. This is because what may appear to be an unconscious or accidental pattern might actually have an intentional mathematical component.

Second, as Eglash examines African designs and knowledge systems, five essential components (recursion, scaling, self-similarity, infinity, and fractional dimension) keep him on track of what does or does not match Fractal Geometry. Since scaling and self-similarity are descriptive characteristics, his first step is to look for the properties in African designs. Once he establishes that theme, he then asks whether or not these concepts have been intentionally applied, and starts to look for the other three essential components. He finds the clearest illustrations of indigenous self-similar designs in African architecture.

The examples of scaling designs Eglash provides vary greatly in purpose, pattern, and method. As he explains, while it is not difficult to invent explanations based on unconscious social forces—for example, the flexibility in conforming designs to material surfaces as expressions of social flexibility—he does not believe that any such explanation can account for its diversity. He finds that from optimization engineering, to modeling organic life, to mapping between different spatial structures, African artisans have developed a wide range of tools, techniques, and design practices based on the conscious application of scaling geometry. Thus, for example, instead of using the Koch curve to generate the branching fractals used to model the lungs and acacia tree, Eglash uses passive lines that are just carried through the iterations without change, in addition to active lines that create a growing tip by the usual recursive replacement.

In his book, *Geometry from Africa: Mathematical and Educational Explorations* (1999), Paulus Gerdes presents the Geometry of the peoples of Africa south of the Sahara that is an outgrowth of their interest in creating and exploring forms and shapes that have blossomed in diverse cultural and social contexts. He provides examples of geometrical ideas in the work of wood and ivory carvers, potters, painters, weavers, and mat and basket makers. He analyzes geometrical ideas in various crafts and explores their educational utility. Employing examples from across Africa, he demonstrates how students may be taught to discover Pythagorean Theorem and to find proofs for it. He also examines connections to Pappus' Theorem, similar right triangles, and Latin and magic squares, in addition to the geometrical ideas in mat and basket weaving, house building, and wall decorations.

While it does not deal exclusively with African Mathematics, the discussion of the subject in George Gheverghese Joseph's *The Crest of the Peacock: Non-European Roots of Mathematics* (2000) is quite extensive and noteworthy. Joseph clearly demonstrates that human beings everywhere have been capable of innovative and advanced mathematical thinking. He traces the history of Mathematics from the Ishango Bone in Central Africa and the Inca quipu of South America about 20,000 years ago to the dawn of modern Mathematics, when the Arabs changed the contours of Algebra around A.D. 830.

What motivated this work is Joseph's desire to correct the widely held belief that Mathematics was essentially a European product. This is because, according to him, the standard treatment of the history of non-European Mathematics is marked by a deep-rooted historiographical bias in the selection and interpretation of facts, and that mathematical activity outside Europe has as a consequence been ignored, devalued or distorted.

In this second edition (the first edition, which is out of print, was published in 1991), Joseph retains the basic format, content and structure of the first one. These are augmented by updates and second thoughts incorporated into notes on each chapter appended at the end in a new section entitled "Reflections."

Joseph's ability to put together such an effulgent work is a reflection of his life story. Born in Karala, southern India, Joseph lived there for nine years. He received his early schooling in Mombasa, Kenya, when his family moved there. He later studied at the University of Leicester, and then worked for six years as an Education Officer in Kenya before returning to the University of Manchester to complete his port-graduate studies. He has held visiting teaching appointments in many parts of the world, including Central and East Africa, Australia, India, New Zealand, Papua New Guinea, and the United States.

The essence of Joseph's masterpiece hinges upon what Stuart Glendinning Hall describes in his thought-provoking essay, "Towards a Working Non-Linear Science of Empowerment" (presented at the Ninth Annual International Conference of the Society for Chaos Theory in Psychology and the Life Sciences in Berkeley, California, July 23-26, 1999). As Hall puts it, the idea that there could be a "people's knowledge" substantially equivalent to "educated knowledge" is acknowledged by Edwin Lazlo in terms of the "growing convergence between the mystical worldview (predominantly, but by no means exclusively, Eastern) and the emerging paradigm of reality among scientists at the cutting edge of contemporary knowledge." It is rare to find anyone arguing this kind of equivalent in the West because, continues Hall, the colonization of consciousness by the values of need and knowledge has had longer to run: "European mathematics is mathematics; all other mathematics is anthropology. That explains why this other mathematics belongs to what has been called ethnoscience" (a la Thomas Crump in his 1990 work, *The Anthropology of Numbers*).

What Joseph accomplishes, therefore, is to highlight the myth in the perception that, again following Hall, "people's collective silence is correlated with stupidity, when it is first and foremost an adaptive response to an environment where people perceive they do not have a voice." Thus, Joseph's book is a valuable collection for every scientist, natural, behavioral, or social, interested in learning about African and other non-European Mathematics.

Jannie van Heerden, Helene Smuts and Chonat Getz in their book, *Africa Meets Africa: Making a Living through the Mathematics of Zulu Design* (2005), and the video that accompanies it explore Zulu cultural tradition and innovation in beautifully crafted objects used everyday. The authors trace the general mathematical ideas on numbers, polygonous figures, tessellations and symmetry in the integrity of the crafts' designs, in addition to the complex styles of weaving, beadwork and pottery employed to make them.

In the first half of the book and the first part of the video, Heerden and his colleagues focus on the historical context and subsequent development of the styles and techniques used by the makers of the craft. In the second half of the book, they explore Zulu design by showing how to think mathematically about particular examples. Due to the integrity of the crafts' designs, the authors discover that general mathematical aspects can be recognized in and taught with particular pieces of beadwork, pots and woven baskets.

All this allows Heerden and his colleagues to offer educators and students from Grade 7 to 12 a familiar context within which they can explore, visually, mathematical ideas on numbers, tessellations and symmetry. In the third part of the video, the authors show how, through various entrepreneurial initiatives, the indigenous knowledge inherent in the crafts—i.e. the ideas and design knowledge that generated them—are finding a place within the free market system.

The preceding discussion clearly shows that the works reviewed here help to shatter long-held myths and misconceptions about Africans, the most pervasive and pernicious of which is the notion of Africans as inactive agents in history. The limitation of these works is that none of the books is comprehensive in its treatment of African Mathematics. What this book does is to bring together the foci of these and other works and to also provide new analyses. Thus, the originality of this book hinges upon the clarity with which familiar but unconnected facts about African Mathematics are marshaled into a simpler and satisfying unity.

ORGANIZATION OF THE REST OF THE STUDY

The rest of this book is divided into 12 chapters. Chapter 2 looks at the mathematical beginnings of bones. Chapter 3 examines Geometry south of the Sahara. Chapter 4 is about numbers. Chapter 5 discusses the beginnings of written Mathematics in Egypt. Chapter 6 narrates the Maghrebian tradition. Chapter 7 investigates Combinatorics. Chapter 8 analyzes Vector Calculus. Chapter 9 reviews the Fourier Transform. Chapter 10 examines Tiling or Tessellation. Chapter 11 discusses Fractals. Chapter 12 analyzes Data Mining. And Chapter 13 presents a general conclusion.

What unifies the chapters in this book can appear rather banal. But many mathematical insights are so obvious, so fundamental, that they are difficult to absorb, appreciate, and express with fresh clarity. Some of the more basic ones will be isolated from accounts of investigators who have earned their contemporaries' respect.

Chapter Two

Beginnings: Mathematics of Bones

INTRODUCTION

This chapter examines the beginnings of Mathematics among Africans: i.e. the Mathematics of Bones. The earliest of this type of mathematical objects in human history include the Lebombo and Ishango Bones that are discussed in this chapter. As Jonas Bogoshi, Kevin Naidoo, and John Webb (1987) point out, the Lebombo Bone, which dates from about 9000 BC, even predates the Czechoslovakian wolf's bone with 57 notches that dates from about 3500 BC.

THE LEBOMBO BONE

In the 1970s, during excavations near what is today called Border Cave in the Lebombo Mountains between South Africa and Swaziland, a small piece of the fibula of a baboon (dated approximately 35000 BC) was found marked with 29 clearly defined notches. At more than 37,000 years old, the bone, which resembles the calendar sticks still in use by the San people in Namibia, ranks with the oldest mathematical objects known (www.math.buffalo.edu/mad/Ancient-Africa). It has also been suggested that since the Lebombo Bone was used as a lunar phase counter, then women may have been the first mathematicians, as keeping track of menstrual cycles requires a lunar calendar (*The Internet Encyclopedia of Science* 2007).

As Richard Manjeiwicz asserts (2001), the Lebombo Bone bears witness to the existence of a very sophisticated accounting system which enabled humans to master time, and it is the first visible hint of the emergence of calculation

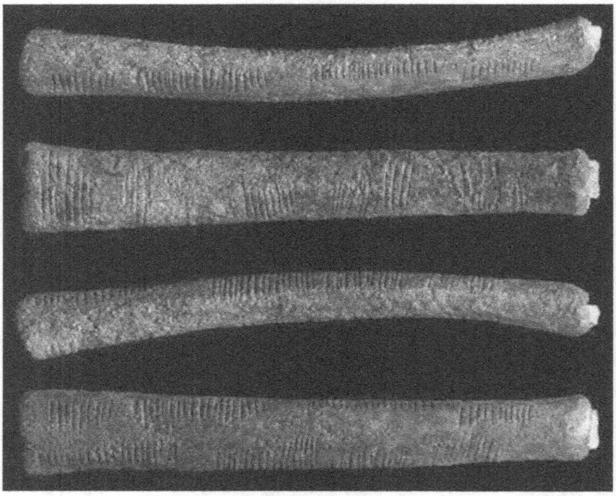

Figure 1. The Lebombo Bone.

in human history. The Lebombo Bone has up to six phases, suggesting that it represents a binary calendar. In the words of Gnaedinger,

> The pattern consists of 14 by 14 dots, linked by two more dots. The 29 dots of the top line can be read as a lunar calendar. There are 30 spaces between and next to the 29 dots. Read the spaces and dots as follows: 30 spaces plus 29 dots plus 30 spaces plus 29 dots plus 30 spaces . . . , yielding 30 29 30 29 30 29 30 29 30 29 30 29 30 29 30 . . . nights or 30 59 89 118 148 177 207 236 266 295 325 354 384 413 443 472 502 . . . nights for 1 2 3 4 5 6 7 8 9 10 11 12 13 14 15 16 17 . . . lunations. There are 2 x 13 x 14 dots below the top line, yielding 364 days; add the dot in the middle of the bottom line and you obtain 365 days for a year (Gnaedinger, 2005).

Furthermore, according to the Kwazulu-Natal Tourism Authority in South Africa, the first known inhabitants of the "Elephant Coast" took residence in the Border Cave, a large overhang in the remote Ingwavuma District, some 200,000 years ago. Some of the oldest evidence setting human evolution in Africa has been found at the Cave, where anatomically modern Homo sapiens remains have been discovered and over a million stone artifacts excavated. Analysis of some of the stone tools has helped scientists to date the introduction of tools crafted into blades and points. In 1942, the Cave yielded the remains of an infant, dating back about 100,000 years, buried in a grave with a shell ornament and red stain suggesting that the body had been painted, pointing to a people capable of abstract and symbolic thought who probably communicated in a fairly complex language. If concern with life after death is taken as a sign of religion, then this is also the oldest record of religion on earth (http://www.southafrica.info/travel/cultural/border-cave.htm).

THE ISHANGO BONE

In Central Equatorial Africa's high mountains, bordering Uganda and the Republic of Congo, is Lake Edward. This small lake, measuring about 60 by 30 miles, is a source of the River Nile. While the area is sparsely populated today, about 6,000 to 9,000 years ago lived a fishing and agricultural settlement by its shores. The community only existed a few hundred years before being buried in a volcanic eruption. The place where the inhabitants' remains were found has a name now given to these people—Ishango. Among their remains discovered in 1960 is the oldest mathematical object, the Ishango Bone, which is now located on the ninth floor of the Royal Institute for Natural Sciences of Belgium in Brussels (de Heinzelin, 1962; Bogoshi et al., 1987).

An examination of the Ishango Bone shows that at one end is a piece of quartz writing. The bone also has a series of notches carved in groups. While it was first thought that the notches are some kind of tally marks as found to

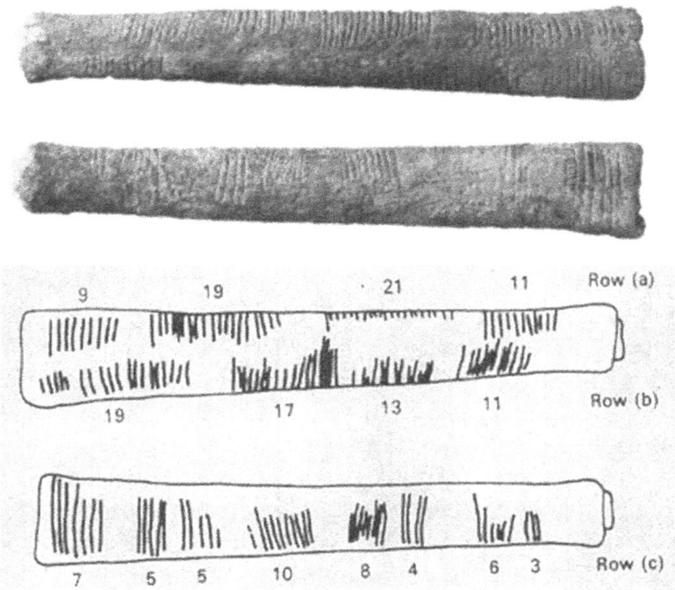

Figure 2. The Ishango Bone.

record counts all over the world, it was later determined that they are much more than a simple tally. The markings on rows (a) and (b) each add up to 60. Row (b) contains the prime numbers between 10 and 20 (11, 13, 17). Row (a) is quite consistent with a numeration system based on 10, as the notches are grouped as 21, 19, 11, 9 = 20 + 1, 20 - 1, 10 + 1, and 10 - 1. Finally, row (c) seems to illustrate the method of duplication used relatively more recently in Egyptian multiplication: eight groups of notches in the order of 7, 5, 5, 10, 8, 4, 6, 3. The last pair (6, 3) is spaced closer together, as are (8, 4), and (5, 5, 10) (de Heinzelin, 1962; Bogoshi et al., 1987).

According to Ron Eglash (1999), the doubling system in the Ishango Bone is fundamental to many of the counting systems of Africa in modern times as well. He adds that it is common, for example, to have the word for an even number 2N mean "N plus N" (e.g., the number 8 in the Shambaa language of Tanzania is *ne na ne*, literally "four and four"). He further mentions that a similar doubling takes place for the precisely articulated system of number hand gestures: for example, "four" represented by two groups of two fingers and "eight" by two groups of four. Andrea L. Petito (1982) finds that doubling was employed in multiplication and division techniques in West Africa, Richard J. Gillings (1972/1982) shows the persistent use of powers of two in Ancient Egyptian mathematics as well, and Claudia Zaslavsky (1973/1999:50) provides archaeological evidence that Ancient Egypt's use of base-2 calculations derived from the use of base-2 in Sub-Saharan Africa.

Recent microscopic investigations illustrate more markings which seem to suggest that the Ishango bone is also a lunar phase counter. Alexander Marshack (1972), for example, conducted a detailed microscopic examination of the bone and found markings of different indentations, shapes and sizes. His conclusion is that there is strong evidence of a close fit between different phases of the moon and the sequential notation contained on the bone, once the additional markings—visible only through the microscope—were taken into account. He further suggests that the different engravings represented markings of various shapes and sizes that may have been a calendar of events of ceremonies or rituals.

For George Gheverghese Joseph (1991/2000), these conjectures about the Ishango Bone highlight three important aspects of proto-mathematics. The first is that the close link between mathematics and astronomy has a long history and is connected to the need felt even among early humans to record the passage of time, out of curiosity as well as practical necessity. The second is that there is no reason to believe that early humans' capacity to reason and conceptualize was any different from that of their modern counterparts. What has changed over time is the nature of the facts and relationships with which humans have to operate. Therefore, the creation of a complicated system of sequential notation based on a lunar calendar was well within the capacity of prehistoric humans, whose desire to keep track of the passage of time and changes in seasons was translated into observations of the changing aspects of the moon. The third is that in the absence of records, conjectures about mathematical pursuits of early humans have to be investigated in the light of their plausibility, the existence of convincing alternative explanations, and the quality of evidence available.

Chapter Three

Geometry South of the Sahara

INTRODUCTION

In this chapter, we discuss a topic in African Mathematics that has yielded a great deal of work: i.e. Geometry south of the Sahara. It is not an overstatement to say that Paulus Gerdes has done the most amount of published work on the topic. The result of his many essays, which are written in English, French, German and Portuguese, is his book titled *Geometry from Africa: Mathematical and Educational Explorations* (1999). Consequently, the discussion in this section draws a great deal from Gerdes' work. Of course, other scholars' works on Geometry south of the Sahara are used and referenced as well. Before exploring these geometrical explorations, we first provide a brief overview of some of the very interesting contributions on the subject.

OVERVIEW OF SOME NOTABLE CONTRIBUTIONS ON GEOMETRY SOUTH OF THE SAHARA

Marcia Ascher in her article, "A Study of Ethnomathematics" (1988), describes the cultural background and mathematical properties of the continuous graphs traced by the Booshung and Tshokwe who live in the Angola-Congo-Zambia region. The Booshung exchange their art for food and raw materials, and classify their designs in interesting ways that are explicated by Ascher. She discusses the problems in continuous tracing among the Bushoong which is primarily the domain of children and the tracing algorithms. She notes that among the Tshokwe, continuously traced graphs play an important role in story-telling. She provides some examples of how some diagrams are utilized to discuss a rite of passage and in connection with the *muyombo* trees that represent the village ancestors. The notion of inside/outside, an aspect dubbed the Jordan curve, is important in some cases. Also discussed are the geometric characteristics of the graphs—for example, many are regular of degree 4—and the algorithms for drawing the curves.

In a trilogy of essays that span ten years, Donald Crowe discusses the Geometry of African art in three societies. In the article, "The Geometry of African Art I: Bakuba Art" (1971), Crowe describes strip and plane patterns in Bakuba art, particularly in textiles and woodcarving. He states that the inspiration for many of the patterns seems to be from weaving, but at least one pattern is a result of sewing together triangles to make bark cloth. He shows that all seven strip patterns and 12 of the 17 possible plane patterns occur. He also describes the relative proportions of some of the patterns and provides an example for each.

Crowe in "The Geometry of African Art II: A Catalog of Benin Patterns" (1975) discusses the strip and plane patterns evident in Benin art. He finds all seven strip patterns and 12 of the 17 frieze patterns to exist. Comparing Benin patterns with those of the Bakuba, Crowe states that glide reflections are rarer in Benin art than in Bakuba art and that Benin patterns are less varied than Bakuba patterns, although the Benin bronze work is unsurpassed. He also provides an example for each of the 12 patterns that occur.

In the article titled "The Geometry of African Art III: The Smoking Pipes of Begho" (1981), Crowe discusses the strip and plane patterns in the art of the smoking pipes of the Krama quarter of Begho in Ghana. He classifies the patterns and discovers that the most common strip pattern is number 7 (referred to as *pmm2* in other systems) and the common plane patterns are *pmm* and *p4m*. According to him, both of these can be easily generated as rows of *pmm3* strips. He also finds all seven strip patterns but only seven of the 17 possible plane patterns to be present.

GEOMETRICAL EXPLORATIONS

According to Paul Gerdes, the development of geometrical thinking started early in African history, as early humans learned to "geometricize" in the context of their labor activities. For example, the hunter-gatherers of the Kalahari Desert in southern Africa learned to track animals, learned to recognize and interpret spoors. They got to know that the shape of the spoor provided information on what animal passed by, how long ago, if it was hungry or not, etc. (Gerdes 1999:3). Such developments propelled Louis Liebenberg to posit that the critical attitude of contemporary Kalahari Desert trackers and the role of critical discussion in tracking suggest that the rationalist tradition of science may well have been practiced by hunter-gatherers long before the advent of the Greek philosophic schools (1990:45).

The major question here, then, is the following: What evidence is available to support these claims? The discussion that follows in the rest of this section is a synopsis of as plenty of the available evidence as possible.

Rock paintings and engravings from all over Africa have been reported (Wilcox 1984). Some of these artifacts date back to several hundreds of years, and others several thousands. They often have geometric structures. Other archaeological finds that indicate geometrical explorations by African hunters, farmers and artisans are stone and metal tools and ceramics. Particularly exceptional are archaeological finds of perishable materials such as baskets, textiles, and wooden objects. The finds from the Tellem are extremely important, as they provide ideas of earlier geometrical explorations (Gerdes 1999:4).

Clear evidence of the exploration of forms, shapes and symmetries exists in the archaeological finds from caves in the Cliff of Bandiagara in the center of Mali. The earliest buildings in the caves are cylindrical granaries made of mud coils that date from the 3rd to the 2nd Century BC. During the 11th Century, the now vanished people of the Tellem (as they are called by the Dogon who inhabit the region from the 16th Century onwards) entered the area from the south, probably from the rain forest. From the 11th up to the 15th Century, the Tellem buried their dead in the remaining old granaries and in new buildings they built in the caves. The dead were buried with wooden headrests, bows, quivers, hoes, musical instruments, baskets, gourds, leather sandals, boots, bags, amulets, woolen and cotton blankets, coifs, tunics, and fiber aprons (Gerdes 1999:6).

These perishable objects found in a reasonably good state of preservation in the caves belong to the oldest objects that have been preserved from Sub-Saharan Africa. Archaeologists and textile experts who have analyzed the Tellem textiles assert that no other region in the world has such a great variety of linear and geometrical patterns in cotton fabrics by means of a single color (the only one available: i.e. indigo). According to Rita Bolland (1991), the Tellem designs have been the object of search for infinite combinations which have persisted to this day.

To illustrate this search by Tellem weavers, Gerdes examines some patterns found on preserved fragments of tunics, sleeves, coifs and caps, woven in plain weave: i.e. the weave in which the horizontal and vertical threads cross each other one over, one under. According to Gerdes, the average width of the threads is 1 mm. The weavers alternated groups of natural white cotton threads with groups of blue, indigo-dyed, threads. From left to right, six vertical white threads are followed by four blue threads; from top to bottom, three horizontal white threads are followed by three blue threads. These yield a plane pattern. The basic rectangle has dimensions ten (= 6 + 4) by six (= 3 + 3), or (6 + 4) × (3 + 3) (Gerdes 1999:6-7).

Gerdes adds that generally, the dimensions are $(m + n) \times (p + q)$, where m, n, p, and q are natural numbers. The Tellem weavers experimented with dimensions and found relationships between the dimensions and the (symmetry) properties of the patterns that resulted. In particular, the variation among the discovered plain weave fragments suggests that the weavers knew the effect on the patterns of the selection of even and odd dimensions, in addition to how these dimensions $(m + n)$ and $(p + q)$ are produced. The Tellem patterns from the 11th and 12th Centuries feature woven rectangles followed by fragments of respective plane patterns, which are two-color patterns in the sense that for each there is a rigid motion of the plane—translation, rotation, reflection—that reverses the blue and white colors (Gerdes 1999:7).

Furthermore, according to Gerdes, the Tellem weavers employed a variant of the plain weave, whereby in one direction double threads are used instead of single threads. In this way, the weavers were able to weave cloths with decorative and strip patterns. With woven cloth, the tailor could begin his/her work: drawing and cutting pieces; knotting, stitching and sewing them together; and decorating, for example, a tunic with a plaited band along the neck opening. Geometric knowledge is imperative in each of these activities. Decorative bands were plaited both with even and odd numbers of strings. Among the plaited bands discovered in the caves, there are on one hand bands made out of 4, 6, 8, and 14 strings, and, on the other hand, out of 5, 7, and 9 strings. The selection of an even or an odd number of strings and the weave, either plain or not, has implications for the visible decorative patterns. In addition, the Tellem weavers also produced blankets made of woolen (Gerdes 1999:9–11).

In his work on textiles from the Kingdom of Kuba in the eastern Kasai region of central Congo, Georges Meurant (1986) demonstrates that the region is one of the African cultural contexts that particularly have stimulated design with a strong geometric accent. The Kuba people are a conglomeration of various culturally related ethnic groups that include the Bushungo, the Ngongo, the Shoowa, and the Ngeende. The reign of King Shyaam a-Mbul (c. 1600–1620) contributed to the intensification of industrious and artistic activities involving copper and iron work, basket and mat weaving, construction of palaces, wood carving, embroidery, pottery and beadwork. Geometric designs decorate all of these objects.

Dorothy Washburn (1990) in her work on Bakuba raffia cloth also in Congo illustrates examples from Bushongo plane patterns. First the men weave the plain light tan raffia cloth on a single heddle loom. The women then employ geometric designs to embroider plush motifs on the plain cloth.

According to Gerdes, Bushongo two-color plane patterns display the structure of woven mats, the decorative pattern on the border of a loincloth, and patterns on women's bark cloth made up of light and dark rectangles, squares, triangles and rhombi sewn together. Among the Ngongo, the decoration of the walls of houses and palaces with mat-work is widespread. The plane patterns entail various symmetries. The sticks are woven together by the vertical lianas. Wooden cups are decorated with symmetrical patterns. An iron knife is decorated with copper in an axial symmetric pattern. Plaid band motifs occur frequently as decoration on wooden and metal objects, carved on gourds, as well as a female Bushoong tattoo. Plaited band motifs made out of only one continuous line are drawn by boys in the sand for amusement. Wood carving of other geometrical line algorithms, leading to strip patterns, exists. Gerdes also mentions Emil Torday and T. A. Joyce's 1910 recording of a "topological" puzzle Torday saw in the Kuba capital. It consists of two pieces of calabash and a string arranged in a way that the player has to separate one of the calabash pieces from the string without cutting or untying the string (Gerdes 1999:13-20).

Donald Crowe (1982) recounts that among the Akan peoples in the forests of present day southern Ghana and eastern Côte d'Ivoire, the Ashanti Kingdom with its capital in Kumasi emerged dominant in the early 18th Century. Before then, from the 14th Century onwards, the city of Begho, just north of the forests, had been a major center of craft and commerce. Among the archaeological finds of Begho are the smoking pipes whose symmetries Crowe analyzed.

According to Gerdes, Ashanti women brought pottery to a high level of perfection by richly decorating them with geometric designs. A jar is decorated with an 8-fold symmetry design and a strip pattern. Other Ashanti cultural activities interwoven with geometrical ideas and experimentation are plaster (wall decoration), smith work (e.g., blades of the state swords) and the weaving and decoration of cloth. To weigh gold dust, the Ashanti have used weights often with animal or geometric forms. Gerdes also cites the first volume of N. Niangoran-Bouah's book, *The Akan World of Gold Weights* (1984), which beautifully displays many of the gold weights with various geometrical shapes and forms (Gerdes 1999:20-23).

Gerdes further notes that among the decorated cloths produced by the Ashanti, is the *adinkra—a* hand-stamped cloth made from white cotton fabric that is divided into rectangles which are filled up by a series of copies of the same stamp. Each stamp is made from a piece of calabash, with sticks glued to its back to serve as a handle. Gerdes also mentions that the best known West African fabric is the multi-colored *kente* cloth that is woven by the Ashanti in Ghana and the Ewe in Togo. *Kente* weavers use horizontal looms to produce long and narrow strips of cloth that are then sewn together to form square or rectangular pieces used to make robes. Gerdes cites Peter Adler and Nicholas Barnard's book, *African Majesty: The Textile Art of the Ashanti and Ewe* (1992), for its beautiful display of examples of *kente* cloth. The examples of the two-color patterns reveal that there exist translations, reflections or rotations which reverse the colors of a particular pattern. Gerdes also cites many other works that provide information about weaving traditions in Sub-Saharan Africa and present colorful photographs of fabrics, many with geometric designs and patterns, examples of which he replicates in his own book (Gerdes 1999:23-24).

Enid Schildkrout and Curtis Keim (1990) point out that in northeastern Congo, the Mangbetu ideal of feminine beauty included body-painting in geometric patterns. Women of the ruling class painted their bodies with *bianga*, a black juice made from the gardenia plant. Mangbetu women wore pieces of bark cloth upon the right shoulder; whenever a woman wanted to sit down, she would place her barkcloth upon the stool before doing so.

Gerdes presents two Mangbetu ceramic water bottles from the 19th Century and examples of useful ivory carvings collected in 1918. He also displays Georg Schweinfurth's 19th Century engraving of the great hall of the court of King Mbuza, giving an idea of its geometrical structure. The hall was used for dancing and has rattles made from strips of palm fibers and filled with seeds used during dances. Gerdes further displays the plane pattern structure of a Mangbetu mural painting photographed before 1915 in Niangara (1999: 29-31).

According to Gerdes, mural decoration is one of the cultural spheres most used for geometrical exploration in Africa. He notes that in West Africa and in southern Africa, it is mostly the women who decorate the walls of their houses with geometrical figures. Each year after the harvest, the women gather to restore and paint their mud dwellings which have been washed away clean by the rains. He also presents beautiful examples of geometrical mural decoration of a house front in Zinder (Niger), mural paintings from the Zande people in northeastern Congo, a Swahili plaster work design from the northern Kenyan coastal region, a Nubian stencil wall painting, a beautifully plaited strip design that decorates part of the wall above the door of the house of a Bamileke chief in Cameroon, a wall decoration motif from Ghana, the roof structure of a Fulani house in Cameroon, a schematic plan of a large Massa farmhouse enclosure in Cameroon, and the circular settlement structure of Zulu cattle keepers in South Africa (Gerdes 1999:31-33).

Also explored by Gerdes are examples of African mat and basket weaving with geometrical exploration. He displays a basket-bowl with ten-fold and a small pendant with five-fold axial symmetry produced by Swazi women using the coiled basket making technique, two coiled basketery motifs from Ovimbundu women in Angola, a Somali coiled basket, a detail of the wall of a plaited basket collected in 1910 in Niangara (noteworthy are the production and joining of woven "toothed squares" and with two-color symmetry), and strip patterns on baskets from the Tabwa people in southeastern Congo (Gerdes 1999:34-36).

Using different materials, designs from mat and basket weaving are frequently applied and further developed in other contexts. Gerdes displays part of the cover of a bottle from Senegal (sections of leather is slit and woven with dried palm fibers), the decoration of a leather money purse from the Mali/Côte d'Ivoire border region and a leather bolster from Korhogo in the latter country, examples of leather appliqués with which Tuareg women in Niger decorate camel saddle bags, a painted strip pattern on a leather bracelet from Tamale in Ghana, a parchment of dyed skin that is richly decorated with geometric designs by female Hausa artisans in Nigeria, several leather shields from the Nandi and Kipsigis peoples in Kenya, a couple of leather shields from the Lukha of Sudan and the Zulu in South Africa, and several others from the Embu and Masai of Kenya and the Sonjo of Tanzania. To make a good sounding drum, it is imperative that the covering is fixed evenly to the drum's wall. Thus, the pins must be equally spaced. Stated mathematically, the holes in which the pins are nailed constitute a regular polygon. Gerdes illustrates this with a drum from Senegal (Gerdes 1999:37-41).

The decoration of calabashes or gourds is another major domain for geometrical expression and exploration. The calabashes are cut open and emptied, and their shells dried in the sun until hard. The yellow surface is darkened or colored, and then designs are applied by incision and/or burning lines into it with a sharp hot metal like a knife or nail. Gerdes presents a design incised and burned onto a semi-spherical calabash from Kano in Nigeria, a burnt-in decoration on a palm wine calabash collected in 1898 in northwestern Cameroon, and some gourd decorations from Nigeria (Gerdes 1999:41).

Wood carving is another activity that summons geometrical exploration. African wooden boxes, seats, headrests, doorposts, pilasters, canoes and boats, spears, drums, pestles, spoons, cups, masks, and combs often have symmetric shapes, in addition to often being decorated with geometric designs. Gerdes displays many examples of these. Various other domains of African life that invite geometrical exploration, imagination and creativity include metalwork, tattooing and other forms of body decoration, hairstyles, string figure and other games (Gerdes 1999:42-47).

The so-called "Pythagorean Theorem" is certainly one of the most interesting and attractive theorems of early geometry. Although Pythagoras of Samos (6th Century BC) is credited with the theorem, the African Egyptologist from

Senegal Cheikh Anta Diop has suggested that Pythagoras may have learned the theorem during his long stay in Egypt (Diop 1980:436, 479). Gerdes and other scholars have shown how diverse African ornaments and artifacts may be used for the discovery of the Pythagorean Theorem and for finding proofs for it. They have also explored connections with related ideas and propositions such as Pappus' Theorem, similar right triangles, Latin and magic squares, and arithmetic modulo *n*.

Gerdes (1999:55-56) observes that in Mozambique, to fasten a cover of a basket, two plaited strips with little lassos at their ends are permanently fixed to the cover, and two cubic buttons are fastened to opposite sides of the basket. Each of the lassos is pulled down around the corresponding button in order to close the basket. Two strips of a palm leaf are used to weave a button. Once an initial knot is made, one continues, on the backside, to weave "one-over-one-under," building up successive layers of the button, until it has become more or less cubic.

One may arrive at the conclusion that

$$C = A + B$$

that is, one arrives at the Pythagorean Theorem. One can also join the square designs to form a pattern in such a way that the oblique lines are continued. Comparing the areas, one may find the following:

$$S + T = U.$$

In African decoration, there are designs or pattern details that display a four-fold symmetry: i.e. a rotational symmetry of 90° occurs frequently. Gerdes explores the idea that four points which correspond to one another under a four-fold symmetry always constitute the vertices of a square (1999:60-68).

Gerdes' first example is a traditional decorative design from Mozambique. He links four corresponding points on the circles by straight line segments and obtains a square. The corresponding points of intersection of these segments with the circles are also the vertices of a square. Instead of linking the four neighboring points of intersection points, Gerdes links the opposed intersection points and obtains a cross that divides the first square into four congruent parts. Both designs produced in this way lead easily to the Pythagorean proposition.

Demonstrably, let a, b and c denote the sides of the congruent right triangles. One may then compute the areas of the squares and triangles and analyze their relationships. As the side of the large square measures $a + b$, its area is $(a + b)^2$. The area of the large square is equal to the sum of the areas of the four right triangles ($4ab/2$) plus the area of the inscribed square (c^2). Therefore

$$(a + b)^2 = 2ab + c^2.$$

Taking into account the equality

$$(a + b)^2 = 2ab + a^2 + b^2,$$

one may find

$$a^2 + b^2 = c^2.$$

In essence, one arrives at the Pythagorean proposition.

Geometrically, Gerdes arrives at this result in other ways as well. For example, he rotates the upper-left right angle clockwise 90° about its lowest vertex. Similarly, he rotates the upper-right triangle counterclockwise 90° about its lowest vertex. He observes that the large square may be considered as composed of one square with side *a*, one square with side *b,* and four right triangles with sides *a, b,* and *c*. Initially, the same large square was composed of a square of side *c* circumscribed by four right triangles with sides *a, b,* and *c*. Therefore, the area of the square with side *c* is equal to the sum of the areas of the squares with sides *a* and *b*.

Gerdes then turns to the cross which divides the initial square into four quadrilaterals that are congruent because of the rotational symmetry. As the arms of the cross constitute the diagonals of a square, they are perpendicular. He let *p* be the side of the initial square and let *r* be the length of the arm of the cross. The four congruent quadrilaterals are joined in such a way that a new and larger square with side *r* appears. At its center, a square hole appears. He let *q* be its side. From the construction of the new square, it immediately follows that

$$r^2 = p^2 + q^2.$$

As $q = a - b$, one sees that *p, q* and *r* make up the side of a right angle. This reasoning can be used to arrive at a "dissection" proof for the Pythagorean proposition. According to E. S. Loomis, this proof was found for the first time (?) by Henry Perigal as late as 1873 (Loomis 1968/1972:104).

Gerdes considers an arbitrary right triangle with sides *p, q* and *r*. He let *p* be larger than *q*. He then draws squares on the legs. Through the center of the square of side p, he constructs a cross with one of its arms parallel to the hypotenuse of the right triangle. The four pieces into which the p-square is dissected are joined together with the q-square to obtain the r-square. Therefore,

$$R^2 = p^2 + q^2.$$

The second example Gerdes provides shows some Chokwe (in Angola) sand drawings with a four-fold symmetry. He easily transforms each drawing into one of the preceding Pythagorean designs.

The third example Gerdes presents is a beautiful tiling from the Kuba people who inhabit the central part of the Congo basin, living in the Savannah south of the dense equatorial forest. The Kuba are famous for their metal products such as weapons and jewelry. Their villages themselves had specialized in the production of certain types of craft like wooden boxes and cups, copper pipes, raffia cloths, etc. Gerdes displays an interesting tessellation of the plane composed of "hooks" and squares, a design that appears traditionally on woven mats and on embroidered raffia textiles, and called *Mikope Ngoma*: i.e. the drums of King Mikope. Each square is embraced by four hooks, with a four-fold symmetry. When the corresponding vertices are linked, a square that has the same areas as the design is obtained.

Gerdes generalizes this design in such a way that the new square still has the same area as the design and that the corresponding right triangle is arbitrary with sides a, b and c. In this way, he finds that the area of the large square (c^2) is equal to the area of the small square plus two times the area of the rectangle formed by joining the hooks. As the small square has side $b - a$, and the rectangles have sides a and $(a/2) + [(b - a) + (a/2)]$, or b, one can conclude that

$$C^2 = 2ab + (b - a)^2 = a^2 + b^2,$$

that is, one proves the Pythagorean Theorem.

The fourth example Gerdes explores is the "elephants' defense" designs of the Kuba in Congo. Gerdes shows two tattooing motifs formerly used by the Ngongo, one of the ethnic groups that belonged to the Kuba Kingdom. When he draws lines between the ends of the tattooing motifs, he obtains a design similar to the Kuba engravings that the Bushongo (the dominant people in the old Kuba Kingdom) call *mwoong*: i.e. "elephants' defense." They call the motif *ikwaakl'imwoong*: i.e. "deformed elephants' defense." When he places several copies of the design together, he discovers, "reinvents" and even proves, easily and geometrically, the Pythagorean Theorem.

When Gerdes employs the identity $(b - a)^2 = b^2 - 2ba + a^2$ and the formulae for the determination of the areas of squares and right angles, he arrives at the same conclusion. The area of the small central square is equal to $(b - a)^2$ and the combined areas of the four neighboring right triangles are $2ab$. Therefore,

(1) $c^2 = (b - a)^2 + 2ab = a^2 + b^2$.

Gerdes offers an ornamental variant and generalization of the Pythagorean Theorem of the Kuba's elephants' defense pattern by employing a big rectangle composed of two pairs of similar triangles and a little rectangle at the center. By placing several copies of this variant together, he discovers/invents and proves the following generalization of the Pythagorean proposition:

(2) $cc' = aa' + bb'$.

He arrives at the same conclusion by algebraic-geometrical reasoning as follows: the area of the central rectangle is equal to $(b' - a)\ (b - a')$ and the areas of the four neighboring right triangles are together $ab + a'b'$. Therefore,

$$c\,c' = (b' - a)(b - a') + a\,b + a'b' = a\,a' + b\,b'.$$

A combination of the Pythagorean proposition with its generalization yields the following: from $c^2 = a^2 + b^2$, $(c')^2 = (a')^2 + (b')^2$, and $cc' = aa' + bb'$, it follows algebraically easily that

(3) $ab' = ba'$,

which implies that

(4) $a : b = a' : b'$.

Stated in another way, the ratios of the sides, taken in the same order, of similar right triangles are equal.

Gerdes returns to the rectangular Kuba variant of the elephants' defense pattern to join some of the ornamental rectangles in order to derive a square. He shows two possibilities of obtaining a square of side $c + c'$. In both cases, a square "hole" appears in the center, whose sides measure $c' - c$. When he extends the sides that end at the vertices of the "hole" until they encounter other sides, four right triangles appear around the square "hole" and, together with the square "hole," they form a new square of side $(a' + b') - (a + b)$.

In both cases, around this new central square emerge four rectangles which comprise patterns that are similar to the Kuba tattooing discussed earlier. Since the large squares are congruent and the small squares are also congruent, Gerdes arrives at the conclusion that both designs themselves are also equal in area. Both are comprised of four rectangles. Therefore, the area (ab') of one of these rectangles of the first design is equal to the area ($a'b$) of one of the rectangles of the second design: that is,

$$ab' = a'b.$$

Stated differently,

$$a : b = a'b.$$

As a result, Gerdes arrives geometrically at the conclusion that the ratios of the sides, taken in the same order, of similar right triangles are equal.

On the basis of the preceding results, it turns out to be easy to prove the Fundamental Theorem of Similar Triangles. Within the same context, Gerdes introduces the concepts of tangent, sine and cosine of an acute angle. Applying the generalization $cc' = aa' + bb'$ in the case of the triangles yields the trigonometric formula

$$c = a\sin\alpha + b\sin\beta.$$

Gerdes moves from widespread decorative motif to a discovery of Pythagoras' and Pappus' theorems and an infinity of proofs. He begins by displaying variations of decorative motif with a long tradition all over Africa. He replicates M.-L. Bastin's examples of the designs of the Chokwe of Angola called *manda a mbaci*, i.e. "tortoise-shield" (Bastin 1961:114, 116), and Alfred Hauenstein's examples of the designs of the Ovimbunda also of Angola referred to as *olombungulu* ("star"), *ononguinguuinini* ("ants"), or *alende* ("clouds") (Hauenstein 1988:39, 50, 54).

Pointing to the number of unit squares on each row of the "star" or "toothed square," Gerdes shows that the area of the "star" is equal to the sum of areas of the 4×4 shaded squares and 3×3 un-shaded squares. A toothed square, especially one with many teeth, looks almost like a real square. The area of a toothed square (T) is equal to the sum of the areas of two smaller squares (A and B):

$$T = A + B.$$

From this, it can be concluded that the area of a toothed square (*T*) is equal to the area of a "real" square (*C*). Since $C = T$, it can be concluded that

$$A + B = C.$$

When Gerdes draws the oblique square (area *C'*) together with the squares with areas *A* and *B*, in such a way that the squares become "neighbors," the Pythagorean proposition for the special case of the right triangle emerges. In essence, toothed squares may assume a heuristic value for the discovery of the Pythagorean proposition.

In his first proof, Gerdes lets *A'* and *B'* be two arbitrary squares. He dissects *A'* into nine little congruent squares and *B'* into 16 congruent squares, and joins the 25 pieces together. The toothed square obtained *T'* is equal in area (*T*) to the sum of the real squares *A'* and *B'*:

$$T = A + B.$$

Once again, the toothed square is easily transformed into a real square *C'* of the same area, allowing one to arrive at

$$A + B = C,$$

i.e. the Pythagorean Theorem in all its generality.

For an infinity of proofs, Gerdes begins by dissecting *A'* and *B'* into n^2 and $(n + 1)^2$ congruent sub-squares for each (positive) integer of *n*. To each value of *n*, there corresponds a proof of the Pythagorean proposition. Put differently, an infinite number of demonstrations of the theorem exist.

The truth of the Pythagorean proposition is almost immediately visible for relatively high values of *n*. One more demonstration of the theorem is delineated when one takes the limit $n \to \infty$. A very short, easily understandable proof is obtained for $n = 1$. Analogously, Pappus' generalization of the Pythagorean Theorem for parallelograms is proved in an infinite number of ways. E. S. Loomis provides 370 different proofs, each requiring its own figure, and he encourages others to find new proofs (1940/1972:13, 269). Gerdes' reflection on a widespread African decorative motif leads him not only to an alternative, active way of introducing the Pythagorean proposition, but also to the generation of an infinite number of proofs of the theorem.

For the toothed squares, Gerdes counts the number of unit squares in each of the horizontal rows. Knowing that the total number of unit squares is a square number, in this case 4^2, he compares the two enumerations and finds:

$$1 + 2 + 3 + 4 + 3 + 2 + 1 = 4^2,$$

or, taking into account the visionary line:

$$(1 + 2 + 3 + 4) + (3 + 2 + 1) = 4^2.$$

It follows that

$$(1 + 2 + 3) + (3 + 2 + 1) = 4^2 - 4,$$

and

$$1 + 2 + 3 = (4^2 - 4)/2,$$

Experimentation with other toothed squares, varying the number of teeth on each side, leads Gerdes to the following extrapolation:

$$1 + 2 + 3 + \ldots + (n - 1) = (4^2 - n)/2,$$

where n denotes a natural number bigger than 1.

By counting all the unit squares of the toothed triangle, row by row, Gerdes finds $1 + 3 + 5 + 7$. Comparing the two enumerations, he concludes that

$$1 + 3 + 5 + 7 = 4^2.$$

Put differently, the sum of the first odd numbers is 4^2. Experimentation with other toothed triangles and comparison of the results leads to the discovery of the general result that the sum of the first n odd numbers is equal to n^2.

Moving from mat weaving patterns to Pythagoras, Latin and magic squares, Gerdes considers a solution found by Chokwe mat makers from northeastern Angola. Each brown strand goes over one white strand and then under four white strands; two successive brown (from the left to the right) differ two in phase. He moves one unit to the right and two units down and calls it the (1, -2)-solution. In this way, the upper part of the detail of a Kuba mat from Congo displays the (1, 2)-solution. Also displayed are the (1, 3) and (1, 4)-solutions. Other solutions are possible if brown squares composed of more than one unit square are admitted.

Gerdes observes that all of these solutions generate tessellations of the plane. What they share in common is that four neighboring brown dots embrace a white square. When he draws straight lines through the centers of these neighboring white squares, a new square grid—an oblique one—is obtained. As each white square is divided by the lines of the new grid into four congruent quadrilaterals and, at the same time, each square of the oblique grid is composed of a brown square at its center, surrounded by four such congruent quadrilaterals, it follows that the area of the oblique square (*C*) is equal to the sum of the areas of one brown (*A*) and one white (*B*) square: $A + B = C$.

When the oblique grid is translated in such a way that its vertices coincide with vertices of the brown squares, one arrives at

(2) $a^2 + b^2 = c^2$,

i.e. at the Theorem of Pythagoras, where *a, b* and *c* denote the lengths of the sides of the squares of area *A, B* and *C*.

A variant is provided by Gerdes whereby the design formed by four neighboring brown dots and the embraced white square shows a rotational symmetry of 90°. The implication here is that any four corresponding points are the vertices of a square. A rearrangement of the white and brown squares leads once again to the Pythagorean Theorem.

Turning to the notion of periodicity and fundamental blocks, Gerdes uses the example of the Chokwe mat, in which the vertical brown strands pass always over one horizontal white strand and then under four white strands; the brown-and-white pattern repeats itself in the vertical direction after five (horizontal) strands. In the horizontal direction, each white strand passes under one brown strand and then over four brown strands. In the horizontal direction, the pattern also repeats itself after five strands. Thus, it can be stated that the pattern has period 5 in both directions: i.e. it can be considered as made up of equally colored 5×5 blocks. This fundamental 5×5 block contains just one little brown square in each row and each column.

To enumerate the unit squares, Gerdes goes from the left to the right, 1, 2, 3, 4, 5 the successive (brown) squares of a fundamental block. He then attributes, to each small square of a specific 5×5 block, a number 1, 2, 3, 4, or 5, in such a manner that it is different from the numbers that appear in the same horizontal, vertical or diagonal line. For the unit square marked by a circle, there is only one possibility: 5.

For a Latin square, Gerdes does the same for the unit squares of the same fundamental block and gets the 5×5 number square. In each row and each column, the numbers 1, 2, 3, 4 and 5 appear precisely once.

Gerdes then uses the Latin square to construct magic squares. A magic square is generally defined as a set of integers in serial order, beginning with 1, arranged in square formation so that the sum total of each row, column and main diagonal is the same. Since the Latin square already has the property that the sum total of each row, column and main diagonal is the same, Gerdes adds 5 to five numbers (one in each row and each column), adds 10 to five other numbers (one in each row and each column), and then adds 15 and 20 in the same way, and obtains a magic square. The sum total of the rows, columns and main diagonal is 65.

These magic squares have additional properties. They are also "diabolic" or "pan-diagonal," meaning that the sum totals of the so-called "broken diagonals" are equal to the same constant 65. "Broken diagonals" refer to those that belong partly to a number square and partly to its equal neighboring number squares. When several equal magic squares of this type are joined, and then displace the 5×5 square, new magic squares are derived.

One of the magic squares has the additional property that each number on one of the four axes of the square, added to the number symmetrically opposite the square's center, gives the same: $12 + 18 + 14 + 16 = 7 + 23 = \ldots = 30$. Reflecting one of the magic squares about its vertical axis, Gerdes gets one of the magic squares presented by Muhammad ibn Muhammad al Fulani al-Kishwani, an early 18th Century Fulani astronomer and mathematician from Katsina (now northern Nigeria), in his 1732 work titled *A Treatise on the Magical Use of the Letters of the Alphabet,* written in Arabic (cited and discussed by Zaslavsky 1973/1999:138–151).

For the Kuba design discussed earlier, Gerdes illustrates that it has the 5 x 5 block as the fundamental block. The alternative solutions have period 10 and period 17, respectively. As done before, by enumerating the successive small squares in the case of the 17 x 17 block yields Latin and magic squares.

Finally, for Arithmetic modulo *n*, Gerdes illustrates that in order to go one place to the right side of the Latin square for the Chokwe mat, he adds 3 modulo 5. In order to go one place up, he adds 1 modulo 5. When he goes one place diagonally down the right, he has to go one place to the right (+3) and then one place vertically downwards (-1); the result is +2. When he moves one place diagonally downwards to the left, he goes, for example, first one place vertically downwards (-1) and then one place left (-3); the end result is -4, but -4 = 1 modulo 5. Going from 2 to 5, he adds 3. When he passes through 4, he adds first 1 and then two times 3, and from 4 to 5, he adds first 3 and then subtracts two times 1. Comparing the two ways yields

$$(1 + 2 \times 3) + (3 + 2 \times [-1]) = 3 \text{ (modulo 5)},$$

or

$$7 + 1 = 3 \text{ (modulo 5)}.$$

The other topics dealing with the Geometry of Sub-Saharan Africa are geometrical ideas in crafts and the *sona* sand drawing tradition of the Chokwe in southern-central Africa. For the former topic, scholars explore symmetrical wall decoration in Lesotho and South Africa; house strip patterns from Guinea, Mozambique, Senegal, and Uganda; and finite geometrical designs from the Lower Congo region. Exploration of a hexagonal basket weaving technique from Cameroon to Kenya, Congo to Madagascar and Mozambique, leads to connections between the underlying geometry and chemical models of recently discovered carbon molecules. Pentagrams are discovered in knots, and alternative ways are developed for rectangle constructions and for the determination of areas of circles and volumes of spatial figures, including a twisted decahedron. For the latter topic, scholars underscore the mathematical underpinnings of the *sona* tradition by developing the Geometry of Lunda-designs and Lunda-patters. A discussion of the intricate mathematics of these topics (hexahedra, octahedra, polyhedra, trigonometric function, Euler's Theorem, monolinearity, algorithms, nonlinearity, etc.) is beyond the scope of this chapter.

Chapter Four

Numbers

INTRODUCTION

Examined in this chapter are African numeration systems. Indeed, the most cited and revered work on this subject is Claudia Zaslavsky's appropriately titled book, *Africa Counts: Numbers and Pattern in African Cultures* (1973/1999). It is only fitting, therefore, that the discussion in this section draws heavily from Zaslavsky's work. Nonetheless, other scholars' works on African numeration systems are also used and referenced accordingly.

NUMERATION SYSTEMS

The economic development of a society, asserts Zaslavsky, ultimately determines the development of its numeration system. African systems of numeration, she adds, varied from a few number words to well-constructed systems with which counting extended into the millions. In many African cultures, a formal system of gesture counting accompanied the number words. Like other peoples around the world, Africans shared a belief in the special significance—for either good or evil—of certain numbers (Zaslavsky 1973/1999:30).

The people naturally required numerals of a higher order as their society's economy grew more sophisticated. In some cases, the people borrowed from neighboring societies; in other cases, neologisms were created. Due to reasons other than the expansion of the numeral system, number words or their meanings undergo alteration over time. As the needs of a society change, special number words associated with the counting of particular objects, such as cowrie currency, tend either to disappear from use or to assume more general application. Furthermore, when a written language is developed, the lengths of various expressions for a given concept in different language varieties are reduced to facilitate standardization in order to spread formal education among the people. For example, the cumbersome Sotho language (Southern Africa) construction for the number 99 changed from *mashome a robileng meno a le mong a (nang le) metso e robileng mono o le mong* ("tens which bend one finger which have units which bend one finger") to *mashome a robong a metso o robong* ("tens nine with a root that is nine") (Zaslavsky 1973/1999:38).

Even more fascinating is how these African numeration systems have survived in the Diaspora. Lorenzo D. Turner (1949/1969), for example, observes that the descendants of enslaved Africans brought to the coastal areas of South Carolina and Georgia during the 18th and 19th Centuries still speak Gullah, a combination of the West African language of their ancestors (Sierra Leone and Liberia) and the English language. Among the Africanisms surviving among the Gullah speakers to the 1940s, according to Turner, were the numerals from one to ten in several West African languages and from one to 19 in the language of the Fulani people. The survival of these number words is a clear indication that the numerals constituted a vital part of the vocabulary of the enslaved Africans who brought their languages from Africa.

According to Zaslavsky, despite the wide distribution of peoples and the existence of about a thousand languages in Africa, the words for two, three and four are similar in an area covering about half of the continent. This area includes the Sudan—the region extending from the Sahara southward to the Gulf of Guinea, and from the Atlantic Ocean to the River Nile—and most of the southern part of the continent now inhabited by Bantu-speaking people. The 300 Bantu languages comprise a subgroup of the Niger-Congo family and bear a great resemblance to the languages of the western Sudanic peoples in terms of basic words. Both categories of languages had a common origin, and half of the Bantu-speaking peoples dispersed throughout the southern half of Africa from the areas of Nigeria and Cameroon (Zaslavsky 1973/1999:39).

Zaslavsky observes that the word for one exhibits great variety in African languages, while this is not the case for the words for two, three and four. The word for two is usually a form of *li* or *di*; the word for three contains the syllable *ta* or *sa*; and the word for four is generally a nasal consonant, like *ne*. The word for five has a variety of forms and is frequently the word for hand.

Zaslavsky quotes Marianne Schmidl (1915:168) who observes the following about the Bantu numeration system:

> When we compare the number words from one to nine in the various Bantu languages, we find a similarity in the names for 2, 3, 4, and 5, while the corresponding gestures differ considerably. The basic stems are:

2 -*vili* or –*vali*
3 –*tatu*
4 –*na*
5 –*tano*

. . . There are various expressions for "one," but generally they are related to -*mwe* (quoted in Zaslavsky 1973/1999:39).

Contrastingly, observes Zaslavsky, there are wide variations in the words for 6, 7, 8 and 9. Thus, she suggests that it is necessary to deal separately with the various branches of the Bantu language when analyzing these numbers (1973/1999:39).

In some cases, the numbers 6, 7, 8 and 9 are expressed as a simple addition to five. For example, in the Kwanyama language (southwestern Africa), one finds the following:

6 *tano-na-mwe*
7 *tano-na-vali*
8 *tano-na-tatu*
9 *tano-na-ne*

In other cases, the composition of one with five to express six may not be so obvious. As Maurice Dalafosse (1929) observes, in the Malinke language spoken on the upper Niger River, six is expressed by *woro* or *wolo*, where *wo* is an abbreviation of the word for five and *ro* or *lo* is a shortened form of one (also cited in Zaslavska 1973/1999:41).

In many African languages, the words for six through nine are derived directly from the gestures of these numbers and are based on several different systems of gesture counting. Interestingly, the Manlike word for nine, *kononto*, means literally "to the one of the belly," a reference to the nine months of pregnancy (Zaslavsky 1973/1999:42).

Beginning with the five-ten system, the Sudanic and Bantu branches of the Niger-Congo language family diverge on the choice of the secondary base. The former generally use 20, and the latter favor ten. Most Bantu languages employ *kumi* or *longo* for ten, albeit it is not the original meaning. Number words for a hundred, a thousand and higher place values were rarely employed, except in association with specific objects to be counted. For example, in the Ziba language,

100 *tsikumu* was used to refer to a string of 100 cowrie shells
1,000 *lukimi* was used to refer to a bundle of ten strings of cowrie currency
10,000 *kakumi* was used to refer to a heap of ten bundles

These same words were applied to other objects, or new words were invented, as the need arose (Zaslavsky 1973/1999:42).

In terms of the five-20 or quinquavigesimal system, for the African languages which build on 20 as a secondary base, ten may be an independent word or it may mean two hands or two fives. This is followed by an expression denoting ten and one, and so on to 15, which may be two hands and one hand or ten and one hand, or even an independent word, as in the Djola language of Guinea-Bissau, where the word means "to bow." Some languages continue to add one, two, etc. to the word for 15 until 20 is reached. The word for 20 in some languages means literally "man complete." In the Banda language of Central Africa, the word for 15 means "three fists" and the word for 20 means "take one person." The same method is repeated by adding to the word for 20 to form the numerals from 21 to 39. The number 40 is expressed as "two men complete," and 100 is "five men complete"—in essence, all of the digits on the hands and feet have been counted five times (Zaslavsky 1973/1999:42–43). And as Delafosse points out, a Malinke expression for 40 is *dibi*, meaning "a mattress," the union of the 40 digits, as the husband and wife lie on the same mattress and have a total of 40 digits between them (1929:389). This is the basic form of the quinquavigesimal system found in many languages of the Sudan region: Djola, Balante and Nalu in the west; Yoruba, Nupe and Efik in Nigeria; Vai and Kru in Liberia and Sierra Leone; and the Nuba language of the eastern Sudan.

As a numeration system develops, special words are either introduced or existing words acquire new meanings. For example, the Malinke word *keme*, meaning "a large number," came to denote 100 and was employed in counting cowrie currency instead of the more cumbersome expression "five men complete" (Zaslavsky 1973/1999:43).

The Igbo (Nigeria) numeration system is based on 20. The number 30 is expressed as the sum of 20 and ten, fifty is 40 and ten, and similarly for the larger numbers. In most language varieties, the formation of the numbers in any place value is by addition of the digits from one to nine to the appropriate word for the multiple of ten (Zaslavsky 1973/1999:43). Roy Abraham (1967) reports an alternative method in the Igbo numerical system: i.e. 16, 17, 18 and 19 can be formed by deducting 4, 3, 2 and 1, respectively, from 20, and similarly for those of the higher place values. The Igbo numeration system also has a special word for the square of 20: 400 is *nu*; the square for 400, or 160,000, is expressed as *nnu khuru nnu*, translated as "400 meets 400." An "uncountably" large number is referred to as *pughu* (Zaslavsky 1973/1999:43).

The Yoruba of southwest Nigeria have a unique system based on 20. It illustrates an unusual subtractive principle still in effect to this day. As

Scott W. Williams points out, the vigesimal (base 20) scale of the Yoruba in Nigeria was present around 1000 AD during the foundation of the Oyo Kingdom. This numerical system hinges heavily upon subtraction. It has different names for the numbers from one (*okan*) to ten (*eewa*). The numbers 11 (*ookanla*) to 14 (*eerinla*) could be translated as "one more than ten" to "four more than ten." But when 15 (*aarundinlogun*) is reached the convention changes, so that 15 to 19 (*ookandinlogun*) are expressed as "20 less five" to "20 less one," respectively, where 20 is *oogun*. Similarly, the numbers 21 to 24 are expressed as subtractions from 30 (*ogbon*). At 35 (*aarundinlogoji*), however, there is a change in the way the first multiple of 20 is referred to: 40 is expressed as "two 20s" (*ogoji*) while 60 (*ogota*), "three 20s," and 80 (*ogerin*), "four 20s," and so on to 200 (*igba*) for "ten 20s." It is the naming of some of the intermediate numbers that the subtraction principle becomes prominent. The following are examples:

45 = (20*3) - 10 - 5
50 = (20*3) - 10
108 = (20*6) - 10 - 2
300 = 20*(20 - 5)
318 = 400 - (20*4) - 2
525 = (200*3) - (20*4) – 2

All the numbers from 200 to 2,000, except those that can be directly related to 400 (*irinwo*), are treated as multiples of 200. From the name *egbewa* (2,000), names are constructed similar to the preceding examples. Indeed, the Yoruba numeral system is quite complicated, as the expression of small terms involves considerable feats of mathematical manipulation (www.math.buffalo.edu/mad/Ancient-Africa).

In terms of linguistic change, the Arabic influence is evident in the Hausa numerical system of northern Nigeria. While 19th Century linguists classified the Hausa numerical system as quinary-vigesimal, but having a base of 12 for the formation of words for 13 through 18, later sources show a clear relationship to the decimal Arabic system for the numeral starting with 20 as well as the Arabic word for six (*shida*). Another aspect of the Hausa numeration system is the use of a subtraction principle for compound numbers ending in eight and nine: for example, 18 = 20 - 2; 19 = 20 - 1 (Zaslavsky 1973:1999:45). W. L. Migeod (1911) reports three designations for 20 in use at the same time in the Hausa numeration system: (1) *hauiya*, a special word for 20 cowrie shells used as currency; (2) the plural of *goma*, the word for ten; and (3) *ashirin*, the Arabic word for 20. The following are the Hausa number words in current use, according to Zalslavsky: 1 *daya*, 2 *biyu*, 3 *uku*, 4 *hudu*,

5 *biyar*, 6 *shida* (from Arabic), 7 *bakwi*, 8 *takwas*, 9 *tara*, 10 *goma*, 19 *goma sha tara*, 20 *ashirin* (from Arabic), 30 *talatin* (from Arabic), and 100 *dari* (1973/1999:45).

There also exist other numerical systems with unusual number bases in Africa. For the Bram and Mankanye people of Guinea-Bissau, five is denoted by the word for "hand," ten is "two hands," nine is "hand and hand less one," and 19 is expressed as "two hands and hand and hand less one," which is definitely a quinary system. But 12 is "six times two." The Bram and Mankanye carry the idea further with 24 = 6 × 4. Their neighbors, the Balante, employ six to delineate the numerals from seven to 12: 7 = 6 + 1, 8 = 6 + 2, etc. For the Ga people of Ghana, both seven and eight are based upon six: 7 = 6 + 1 and 8 = 6 + 2 (Zaslavzky 1973/1999:46).

The Huku of Central Africa display the greatest variety, as they base their numerical system on four and six (Zaslavsky 1973:1999:46):

7 = 6 + 1
8 = 6 + 2
9 = (2 × 4) + 1
10 is an independent word
13 = 12 + 1
16 = (2 × 4) × 2
17 = (2 × 4) × 2 + 1
20 = 2 × 10

In his doctoral dissertation, "Number Systems and Calendars of the Berber Populations of Grand Canary and Tenerife in the 14th-15th Centuries" (1997), José Barrios Garcia discusses how Grand Canary and Tenerife were inhabited in 14th and 15th Centuries by Berber populations called Canarians in Grand Canary and Guanches in Tenerife and the numeration and Calendrical systems they developed on the islands. The population on each island was between 40-60,000 inhabitants, sustaining a notably developed agricultural (barley and wheat) and cattle raising (goats, sheep and pigs) economy. The islands were incorporated to the Spanish crown in the late 15th Century

According to Garcia, the economic characteristics on the islands required a certain number of arithmetical and calendrical activities that available written ethnographic sources have corroborated. His analysis of the preserved notes of the number systems led him to delineate the following numerical system that was being used: a pure 10-based system that was deeply related to both proto-Berber and ancient Egyptian numeral systems, without discarding a possible concurrent use of a 12-based system related with calendrical counts, as well as the existence of a systematic census of the inhabitants of each island. Garcia also establishes the existence of systematic records of lunar, solar and sidereal counts (Garcia, 1997).

Garcia complimented his research of Grand Canary with an Archaeoastronomical study of the mountain of Cuarto Puertas, considered as having great religious importance. From the evidence he collected, Garcia infers that at the top of the mountain is a summer solstice marker that works by means of the shadow a certain rock casts at sunrise upon a great sign carefully carved on the opposite wall. Also, archaeological, ethnographical and linguistics evidence Garcia collected led him to suggest that the Canarians systematically recorded numerical, astronomical and calendrical data by means of geometrical figures (squares, triangles, circles, etc.) painted in white, red and black on wood planks and on the walls of certain caves. From the evidence he gathered from the decoration of the Painted Cave of Galdar, Garcia states that the inhabitants used a chessboard of 3 (vertical) × 4 (horizontal) squares, named acano, to represent 12 moons. On this basis, Garcia suggests that acano is a lunar calendar and shows how the vertical numeration of its squares forces the moons of solstice, equinox and eclipse to move across the board with very simple and stable patterns. Garcia asserts that these patterns provide a safe and clear mnemonic guide for performing on the acano an easy arithmetical calculus of seasonal and eclipse moons over extended periods by using the difference in days of the lunar year with either the solar year or the eclipse year to perform an elementary saw function on the squares (Garcia, 1997).

According to Garcia, this calculus establishes the octaeteris and the 135-moon eclipse cycle as basic periods of the acano. Since the Canarians observed the summer solstice and had important festivals on the accompanying crescent moon, Garcia examines two notes from ancient written records that support the postulate that the inhabitants measured a one and half eclipse year as 520 days. For Garcia, this proposed calculus on the acano reveals an unsuspectedly high level of Canarian mathematical astronomy (Garcia, 1997).

In contrast with some notes supporting the existence of a Sirius calendar in Grand Canary, the Gaunche calendar played a fundamental role in the former by the phases of the star Canopus. The helical rise around the middle of August fixed the first moon of the Gaunche lunar calendar, while its helical set in late April and its acronical rise in late January fixed the two other well documented feasts of the island. The Ggaunche cult to this star became what emerged as the main Catholic cult of the island after the Spanish conquest: i.e. the Virgin of Candelaria. Additional evidence gathered from continental Berbers supports the antiquity and widespread of the Canopus cosmological system in Northwest Africa (Garcia, 1997).

In terms of record keeping among the Gaunche, Garcia notes that written sources state the utilization of tally woods, particularly clay beads joined with a string to form a sort of necklace, usually found in funerary Gaunche caves. But as he cautions, more reliable archaeological evidence is needed to corroborate the written notes (Garcia, 1997).

Other aspects of African numerical systems that have been thoroughly investigated include gesture counting, taboos and mysticisms, the concept of time, numbers and money, weights and measures, games, magic squares, patterns and shapes, and histories of various numerical systems. Due to their detailed nature, however, these aspects are not discussed in this chapter.

Chapter Five

Beginnings of Written Mathematics: Egypt

INTRODUCTION

We discuss in this chapter a topic in African Mathematics that has yielded a great deal of work: i.e. the beginnings of written Mathematics in Egypt. Before exploring the existing findings, we first provide a brief overview of some of the very interesting contributions on the topic.

OVERVIEW OF SOME NOTABLE CONTRIBUTIONS ON THE BEGINNINGS OF WRITTEN MATHEMATICS IN EGYPT

N. L. Briggs in his essay, "The Roots of Combinatorics" (1979), explains that the most ancient mathematical problem connected with Combinatorics may be Problem 79—i.e. the house-cat-mice-wheat problem—of the Ahmes or Rhind Papyrus. He adds that this mathematical problem appears in a similar form in a problem of Fibonacci's *Liber Abaci* and in an English nursery rhyme.

In his article titled "Egyptian Arithmetic" (1981), Evert Bruins discusses how he used an extensive computer search to analyze the construction of the $2/n$ table in the Ahmes Papyrus. While he does not provide a definitive answer, he does raise doubts about earlier theories on the aspect.

Nathan Altshiller-Court in "The Dawn of Demonstrative Geometry" (1964) argues that it appears unlikely that the Greeks could have invented their notion of proof so rapidly and in isolation. This is because, according to him, the notion of geometric proof was a secret that was jealously guarded by those in the "inner sanctum" of the Egyptian priesthood and kept away from everyone else.

In his article titled "Quadrature of the Circle in Ancient Egypt" (1977), Hermann Engels describes the Egyptian formula for the area of a circle in terms of the practices of Egyptian stone masons. He explains how the Egyptian masons covered their designs with a grid in order to form a relief.

Howard Eves' "On the Practicality of the Rule of False Position" (1958) discusses how the method of false proposition can be simpler than similar contemporary methods by giving an example from the Ahmes Papyrus. Eves also likens the rule of false position to the method of similitude in geometric constructions.

"The Area of the Curved Surface of a Hemisphere in Ancient Egypt" (1970) by E. N. R. Fletcher discusses Problem 10 of the Moscow Papyrus on the surface area of a basket and is perceived by some mathematicians to compute the surface area of a hemisphere. Fletcher examines which units might have been utilized in the problem and proffers the postulate that the basket in question was in reality hemispherical and was designed to hold 100 *Hekat* of corn. Furthermore, he points out that the units utilized in ancient Egypt appear to have some interesting geometrical properties. An example is that a circle with a radius of 1 *pes* (or 1 "foot" = 16 digits) was approximately equal in area to a square with sides measuring 1 royal cubit.

In two essays, R. J. Gillings, discusses two mathematical aspects of the Ahmes Papyrus. In the artcle titled "Problems 1 to 6 of the Rhind Mathematical Papyrus" (1962), Gillings discusses how 1, 2, 6, 7, 9 loaves of bread are divided among 10 men. The results are given as unit fractions, including ⅔ as a unit fraction. In "The Volume of a Truncated Pyramid in Ancient Egyptian Papyri" (1964), Gillings provides a method to derive the formula $V = ⅓(a^2 + ab + b^2)$ for the volume of a truncated pyramid, employing only the formula for the volume of a complete pyramid and other methods that the Egyptians had utilized. He also shows how fairly simple arguments sufficed when $b = a/2, a/3, \ldots$, and also when $b = ⅔$a.

In her book chapter titled "Egyptian Mathematics" (2007), Annette Imhausen presents a selection of sources and discusses the characteristic features of Egyptian Mathematics. Since the selection is taken from over 3,000 years of history, Imhausen employs examples from specific contexts. She shows that the mathematical texts provide different types of mathematical problems and that the majority of the problems deal with administrative matters.

Beatrice Lumpkin in her "Note: The Egyptians and Pythagorean Triples" (1980) shows that some ancient Egyptian problems suggest a knowledge of certain Pythagorean triangles. She cites an example in the Berlin Papyrus in which there are problems where a given square is to be written as the sum of two squares in a given ratio. The solutions are predicated on the fact that $6^2 + 8^2 = 10^2$. These facts are akin to the contemporary knowledge of (3, 4, 5) right triangle. Lumpkin also discusses the fact that the Egyptian units of measurement suggest knowledge of the Pythagorean Theorem in the special case of an isosceles right triangle. The *double remen* equals the diagonal of a square whose side equals one *cubit*. Changing the units of measurement from cubits to *double remens* would double the area of a figure.

Finally, in his article, "Egyptian Fractions" (1981), Charles Rees uses the Egyptian preference for dealing with unit fractions, with the exception of the case of $^2/_3$, as a starting point for certain problems in Number Theory. He provides several proofs to show that every fraction can be represented as a sum of unit fractions that vary in the number of fractions produced and the maximum size of the denominators. He also discusses various conjectures about unit fractions. For instance, $4/n$ and $5/n$ can always be written as the sum of three or less Egyptian fractions. He further generates some interesting results such as the denominators of the unit fractions being required to be squares, be cubes, or be square free.

WRITTEN EGYPTIAN MATHEMATICS

According to Williams, as early as 4800 BC, the Egyptians had developed a calendar. In 4200 BC, their Mathematics and Astronomy produced a 365-day calendar (12 months of 30 days plus five feast days). By the time of the Egyptian mace in 3100 BC, various agricultural communities along the banks of the River Nile were united by a Nubian, King Menes, who founded a dynasty of 32 Pharaohs and lasted 3,000 years. For more than half that time, areas of modern Israel and Syria as well as the Nile Valley were part of Egypt. To rule effectively, the Egyptians had to develop an extensive and efficient administration for collecting taxes, taking census, and maintaining a large army. Since all of these activities required Mathematics, the Egyptians at first employed counting glyphs. But even by 2000 BC, the hieratic glyphs were being used by Egyptians (www.math.boffalo.edu/mad/Ancient-Africa).

Williams adds that from Herodotus, the Greek (-500 BC), one learns that the Pharaoh Ramses II (-1300 BC) divided the land into lots and gave each man a square piece of equal size, from the produce of which he exacted an annual tax. If a man reported that his holding was damaged by the encroachment of the river, the Pharaoh would send inspectors and surveyors to measure the extent of the damage to assess the extent of the loss. The data would then be used to set a fair proportion of his property tax. This may be the way in which Geometry was invented and subsequently reached Greece (www.math. boffalo.edu/ mad/Ancient-Africa).

Furthermore, Williams notes that another Greek, Democritus (-410 BC), while boasting, made a high estimation of Egyptian Mathematics. According to him, no one surpassed him in the construction of lines with the help of a ruler and compass, not even the so-called "rope stretchers" (surveyors) among the Egyptians (www.math.boffalo.edu/mad/Ancient-Africa).

As stated earlier, among the many strands of African Mathematics, the Egyptian one has been studied and written about the most. Among the topics dealing with Egyptian Mathematics that have been investigated are the sources, number recording, Arithmetic, Algebra, and Geometry. Fortunately for us, George Gheverghese Joseph has discussed all of these aspects in his book titled *The Crest of the Peacock: Non-European Roots of Mathematics* (1991/2000). What follows in the rest of this section is a synopsis of his presentation. (It behooves me to note here that Annette Imhausen's book chapter titled "Egyptian Mathematics" [2007] is a valuable contribution, albeit her treatment of the subject is mostly in terms of the administrative uses of the Mathematics and she fails to mention Joseph's work.)

Time has not been kind to Egyptian mathematical sources recorded on papyri. Nonetheless, the papyrus is quite a bit more durable than the palm leaves, bark and bamboo used in writing materials by the ancient Chinese and Indians. There are two major sources and a number of minor ones on Egyptian Mathematics. A majority of the minor sources preserve the Mathematics of later epochs of Egyptian history: the Hellenistic (332–39 BC) and the Roman (30 BC–385 AD). The most important major source is the Ahmes (or Ahmose) Papyrus, named after the scribe who composed it, in about 1650 BC, from a work three centuries older. It is also refereed to as the Rhind Mathematical Papyrus, after the British collector who acquired it in 1858 and subsequently donated it to the British Museum. The second major source is the Moscow Papyrus written around 1850 BC; it was taken to Russia in the middle of the 20th Century, ending up in Moscow's Museum of Fine Arts. Between the two papyri, there are 112 mathematical problems with their solutions (Joseph 1991/2000:59–60).

Among the minor sources, there is the Egyptian mace head from the early third millennium BC which contains the war booty of the Pharaoh Namer: 120,000 prisoners; 400,000 oxen; and 1,422,000 goats. Recording these numbers required a highly developed system of numerals that allowed counting to continue indefinitely by the introduction of a new system whenever necessary. Other minor sources include the Egyptian Leather Roll, a table consisting of 26 decompositions into unit fractions, from the same period as the Ahmes Papyrus; the Berlin Papyrus which contains two problems of simultaneous equations, one of second degree; the Reisner Papyri, consisting of four worm-eaten rolls which record volume calculations relating to temples, from around 1800 BC; and the Kahun Papyrus, containing six scattered mathematical fragments, not all of which have been deciphered, from around 1800 BC (Joseph 1991/2000:60).

According to Ahmes, his material came from an earlier document belonging to the Middle Kingdom (2000–1800 BC), and he suggests that the knowledge may ultimately have been derived from Imhotep (c. 2650 BC), the legendary architect and physician to Pharaoh Zoser of the Third Dynasty. The opening sentence of the Ahmes Papyrus is a testament that it is "a thorough study of all things, insight into all that exists, knowledge of all the obscure secrets" (Joseph 1991/2000:60–61). It contains 87 mathematical problems and their solutions, making it the most comprehensive source of early Egyptian Mathematics. There are 25 problems, among them two notable results of Egyptian Mathematics: (1) the formula for the truncated square pyramid (or frustum) and (2) a remarkable solution to the problem of finding the curved surface area of a hemisphere (Joseph 1991/2000:61).

The dominant impression in many textbooks on the history of Mathematics that only one scheme of numeration, the hieroglyphic (pictorial), was employed in ancient Egypt is incorrect. Instead, at least three different notational systems existed: (1) hieroglyphic, (2) hieratic (symbolic), and (c) demotic (popular). The first two numeration systems appeared quite early in Egyptian history. The hieratic notation was used in both the Ahmes and Moscow Papyri, while the demotic variant was a popular adaptation of the hieratic notation and became important during the Greek and Roman periods of Egyptian history (Joseph 1991/2000:61).

In the hieroglyphic system, characters represented objects, some easily recognizable. Special symbols were employed to represent each power of 10 from 1 to 10^7. A unit was therefore commonly written as a single vertical stroke; when written in detail, however, it was shown as a short piece of rope. A long piece of rope in the shape of a horseshoe was used to represent the number ten, and a coil of rope was employed for 100. Quite in keeping with the important role of the "rope stretchers" (i.e. surveyors) in ancient Egypt, the overriding motif in all of these cases seems to have been a rope whose length and shape determined the magnitude of the number represented. A thousand was represented by a pictograph that resembles a lotus flower, although the plant sign formed the initial of *khaa*, which means cord. The number 10,000 was represented by a crooked finger; 100,000 was represented by a stylized tadpole; 1,000,000 was represented by a man with arms upraised; and 10,000,000 was represented by a rising sun. Any number can be written by using these symbols additively (Joseph 1991/2000:61–62).

The hieroglyphs were arranged in descending order of magnitude from right to left. To add two numbers, an individual made a collection of each set of symbols that appeared in both numbers and replaced them with the next higher symbol as necessary. To subtract, the process for addition was reversed and the higher hieroglyph is replaced by the lower one accordingly (Joseph 1991/2000:62).

The hieratic representation was similar to the hieroglyphic process, since it (i.e. the hieratic process) was additive and based on powers of ten. But the former was far more economical, as a number of identical hieroglyphs were replaced with fewer symbols, or just one symbol. For example, 999 would require 27 symbols in hieroglyphs compared to six in hieratic. While this process was more taxing on memory, its economy, speed, conciseness and greater suitability for writing with pen and ink must have precipitated its early adoption in ancient Egypt. From the point of view of the history of Mathematics, the hieratic system may have inspired the development of the alphabetic Greek number system around the middle of the first millennium BC (Joseph 1991/2000:62-63).

A great feat of the Egyptian method of duplication or division is that it requires prior knowledge of only addition and the two times table. Every integer can be expressed as the sum of integral powers. Thus,

$$15 = 2^0 + 2^1 + 2^2 + 2^3$$
$$23 = 2^0 + 2^1 + 2^2 + 2^4$$

The confidence with which the ancient Egyptians approached all forms of multiplication by this process suggests that they were aware of this general rule (Joseph 1991/2000:63-64).

This ancient method of multiplication was the foundation for Egyptian calculation. The Greeks employed it with some modification and it continued to be used by Europeans well into the Middle Ages. A modern variation of the method is still popular among rural communities in Russia, Ethiopia and the Near East (Joseph 1991/2000:65).

The ancient Egyptian process of division was closely related to the method of multiplication. For example, a division *x/y* is introduced by the words "reckon with *y* so as to obtain *x*." So for the Egyptian scribe, instead of thinking of "dividing 696 by 29," he would say to himself "Starting with 29, how many times should I add it to itself to get 696?" If a scribe were to encounter a problem of not being able to get any combination of the numbers in the right-hand column to add up to the value of the dividend, he would introduce fractions. In essence, the writing method of the ancient Egyptians did not allow any unambiguous way of expressing fractions; consequently, they tackled the problem in quite an ingenious way (Joseph 1991/2000:65–66).

Operating with unit fractions is a singular feature of Egyptian Mathematics that cannot be found in any other mathematical tradition. A large proportion of ancient Egyptian calculations utilized such operations. The great emphasis on fractions hinged on two reasons. The first reason is that since the society did not use money, transactions having to be carried out in kind, there was a need for accurate calculations with fractions, particularly in practical problems such as division of food, parceling out land, and mixing different ingredients for beer and bread. The second reason emerged from the particular character of Egyptian Arithmetic: i.e. the process of halving in division often led to fractions (Joseph 1991/2000:66–67).

The utility of unit fractions in arithmetical operations and the peculiar system of multiplication yielded a third aspect of Egyptian computation: i.e. every multiplication and division involving unit fractions invariably led to the problem of how to double unit fractions. Doubling a unit fraction with an even denominator is a simple matter of halving the denominator. Therefore, doubling 1/2, 1/4, 1/6 and 1/8 yields 1, 1/2, 1/3 and 1/4. Doubling 1/3 raised no difficulty, since 2/3 had its own hieroglyphic or hieratic symbol. Difficulty arose when doubling unit fractions with other odd denominations. Since it was difficult in Egyptian computation to write two times $1/n$ as $1/n + 1/n$, the need arose to develop a table that provided the appropriate unit fractions that summed to $2/n$, where $n = 5, 7, 9, \ldots$. The major purpose of constructing the $2/n$ table was to use it for multiplication and division (Joseph 1991/2000:68).

In order to deal with "problems of completion," the ancient Egyptians developed a method known as "red auxiliaries" (so named because the scribes wrote these numbers in red ink). The approach is analogous, but not equivalent, to the present-day method of least common denominator. First, the Egyptians took the denominator of the smallest unit fraction as a reference number and then multiplied each of the fractions by this number to obtain "red auxiliaries." They proceeded to calculate by how much the sum of these auxiliaries fell short of the reference number. This shortfall quantity was then expressed as a fraction of the reference number to get the complement. If the shortfall quantity turned out to be an awkward fraction, a further search was done for a reference number that would result in more manageable auxiliaries (Joseph 1991/2000:71–72).

Rules devised by mathematicians for solving problems about numbers of one kind or another can be classified into three types. In the early stages, the ancient Egyptian rhetorical type, which gave birth to the two that followed, developed rules that were expressed verbally and consisted of detailed instructions about what was to be done to obtain the solution to a problem. In time, the prose of this rhetorical form of algebra paved the way for the use of abbreviations for recurring quantities and operations, ushering the appearance of "syncopated algebra." Traces of this algebra can be found in the works of the Alexandrian mathematician Diophantus (c. 250 AD). During the past 500 years, there has developed "symbolic algebra" by which, with the aid of letters and signs of operation and relation (+, - , x, ÷, =), problems are stated in such a form that the rules of solution may be applied consistently and systematically (Joseph 1991/2000:76).

A series is the sum of a sequence of terms, and the most common types are the arithmetic and geometric series. In an arithmetic progression (AP), each term after the first (usually denoted by *a*) is obtained by adding a fixed number, called the common difference (usually denoted by *d*), to the preceding term. For example, 1, 2, 3, 5, 7, 9, . . . is an AP with $a = 1$ and $d = 2$. For a geometric progression (GP), each term after the first (*a*) is formed from the preceding term by multiplying by a fixed number called the common ratio (usually denoted by *r*). For example, 1, 2, 4, 8, 16, . . . is a GP with $a = 1$ and $r = 2$. The ancient Egyptian method was based on operations with the basic GP 1, 2, 4, 8, . . . and an understanding that any multiplier may be expressed as the sum of elements of the sequence (Joseph 1991/2000:79).

When it comes to the great achievements of Egyptian Geometry, there is general agreement on two: (1) the approximation to the area of the circle and (2) the derivation of the rule for calculating the volume of a truncated pyramid. Some disagreement arises over the third: the correct formula for the surfaced area of a hemisphere.

In the Egyptian method of calculating the area of a circle, the implicit estimate of ϖ can be worked out quite easily by equating A with A^E:

$$\pi d^2/4 = (8d/9)^2$$

from which the following is obtained:

$$\pi = 4(8/9)^2 = 256/81 = (16/9)^2 \sim 3.1605$$

This is nearly the value that is obtained by taking $d = 9$ in the expression $A^E = (8d/9)^2$ (Joseph 1991/2000:82–83).

The Egyptian knowledge of the correct formula for the volume of a truncated square pyramid is the zenith of Egyptian Geometry. The Egyptian approach is correctly based on the following formula:

$$V = (a^2 + ab + b^2)h/3$$

Three major explanations have been advanced for how the Egyptians arrived at the correct formula for the volume of a truncated pyramid. The first explanation is that the truncated pyramid was cut up into smaller and simpler solids whose volumes were then estimated before putting them back together. The reduction of the sum of the volumes of all the component solids to the final formula would require a very high degree of algebraic knowledge and sophistication (Joseph 1991/2000:87).

The second explanation is that the Egyptians had discovered empirically that the volume of a truncated pyramid can be derived as the product of the height of the frustum, h, and the Heronian mean (the Heronian mean of two positive numbers x and y is given by $(x + y + \sqrt{xy})/3$) of the areas of the bases a^2 and b^2. This explanation is supported by the Alexandrian mathematician Heron (or Hero) of the 1st Century AD, whose work, *Metrica*, contains a useful synthesis of Egyptian, Greek and Babylonian traditions. Book II of the work details the volume calculations of prisms, pyramids, cones, parallelepipeds and other solids. The method for calculating the volume of a truncated pyramid using the mean derived directly from the Egyptian mathematical tradition (Joseph 1991/2000:87).

The final explanation is that the volume of a truncated pyramid was calculated as the difference between an original complete pyramid and a smaller one removed from its top. This explanation has been said to be the most plausible of the three, given the concrete approach to Geometry favored by the Egyptians. Regardless of how the Egyptians came to the discovery of this formula, it remains to this day a lasting testimony of their mathematical ingenuity (Joseph 1991/2000:87).

Chapter Six

The Maghrebian Tradition

INTRODUCTION

Thanks to Ahmed Djebbar (1995), there exists now a comprehensive survey of the mathematical tradition of the Maghreb (North Africa) from the 9th to the 18th Century. This chapter is a summary of Djebbar's work.

From the 9th to the 11th Century, which came to be known as the period of the installation and consolidation of Muslim power in the first cities of Spain and the Maghreb, Medicine and Calculation were the first scientific endeavors to benefit from teaching, followed by the publication of works, to respond to the needs of certain higher-rank people of society and to lawyers for the resolution of certain problems such as those involved in land measurement or in the partitioning of inheritance. The eminent role played by Kairouan (Holy City in northern Tunisia) in the theological debates during the Aghlabid epoch (800-910) attracted numerous intellectuals from the East to Ifriqiya (city in today's western Libya), among them men of science educated in arithmetical and geometrical techniques used to solve problems of land measurement and inheritance. As in other regions of the Islamic world, the patronage in favor of scientific activities existed in the Maghreb between the 9th and 11th Century and functioned in the image of that of the great metropolis of the East: buying books, financing production of manuscripts, remuneration for scholars, and construction of schools and institutions (Djebbar, 1995).

THE GIANTS OF MAGHREBIAN MATHEMATICS

Very little is known about the mathematical activities of the Maghreb during the 10th Century. It appears that the patronage system started by the

Aghlabids in the 9th Century was continued and benefited Mathematics and Astronomy, especially during the first two decades of the government of the Fatimid caliph al-Mucizz (953–975). Specialists often underlined the importance of the political and economic history of 12th Century Maghreb. Therefore, it is reasonable to suggest that their importance extended to other spheres of social life as well. The cultural and scientific history of this era, however, remains vastly unexplored (Djebbar, 1995).

One of the most outstanding mathematicians of the Maghrebian tradition was Abu al-Qasim al-Qurash. He was a specialist in Algebra and in the Science of Inheritance, in addition to his expertise in certain religious sciences. In Algebra, al-Qurash is known for his commentary on the book of the great Egyptian mathematician Abu Kamil (d. 930). Its importance is confirmed by the historian Ibn Khaldun (d. 1406) who considered it as one of the best treatises written on the book. This work of al-Qurash was not a simple commentary on a famous treatise of the Algebra of its time. Indeed, some new elements were introduced in the commentary. Beginning at the level of presentation, al-Qurash provides the objects and the operations of Algebra before explaining the solution of the canonical equations followed by the demonstration of the existence of the solutions of these equations. Al-Qurash distinguished himself from his predecessors in the classification of six canonical equations and in the demonstrations. This work continued to be studied and taught up till the 14th Century (Djebbar, 1995).

In the domain of inheritance, al-Qurash is known for having elaborated a new method based on the decomposition of the numbers in prime factors in order to reduce the fractions that intervene in the distribution of a given inheritance to the same denominator. His approach was quickly appreciated by mathematicians who wrote the handbooks explaining the method and showing its utility through the presentation of concrete problems of inheritance (Djebbar, 1995).

Another outstanding mathematician of the Maghrebian tradition was al-Hassa. This mathematician was also well-known as a reader of the Qur'an, as an inheritance specialist, and as a high-ranking citizen since he was given the title of *Shaykh al-Jamaa* (Chief of the Community). He lived for some time at Sebta (in the Extreme Maghreb) with other mathematicians of this city (Djebbar, 1995).

Two of al-Hassa's books have survived. The first book titled *Book of Proof and Recall* is a handbook of calculation treating numeration, arithmetical operations on whole numbers and on fractions, extraction of the exact or approximate square root of a whole of fractionary number and summation of progressions of whole numbers (natural, even or odd), and of their squares and cubes. The book occupies an important place in the history of mathematics for the following three reasons: (1) the manual remains the most ancient work of calculation representing simultaneously the tradition of the Maghreb and that of Muslim Spain; (2) it is wherein one finds a symbolic writing of fractions, which utilizes the horizontal bar and the dust ciphers—i.e. the ancestors of the digits that we use today; it is the only Maghrebian work of calculation known to have circulated in the scientific foyers of south Europe, as Moses Ibn Tibbon realized, in 1271, a Hebrew translation of it (Djebbar, 1995).

The second work is titled *The Complete Book on the Art of Number*. Only the first part of this book that contains 117 folios was recovered and identified in 1986. Its content takes up themes in the first book dealing with whole numbers, by developing them, and presents new chapters on the decomposition of a number in prime factors, the common divisors and multiples, and the extraction of the exact cubic root of a whole number. The second part of the work, which has not been recovered, is said to be dedicated to operations of fractions; the summation of the different categories of whole numbers; and the exposition of the algorithms that facilitate the calculation of perfect, deficient, abundant, and amicable numbers (Djebbar, 1995).

A third member of the Maghrebian mathematical tradition of the 12th Century is Ibn al-Yasamin, who is relatively better known than the preceding two. Al-Yasamin's mother was a Black African and his father was a Berber. For a long time, this mathematician was only known for his minor work of 62 lines, *Poem on Algebra*. It was the success of this poem that prompted him to write a similar one that deals with the roots of numbers and another one that summarizes the method of false position (Djebbar, 1995).

Al-Yasamin's work titled *Fecundation of the Spirits with the Symbols of the Dust Ciphers* is much more important than the three poems, at both the quantitative and qualitative levels. It is a book of more than 200 folios that deal equally with the classical chapters of the Sciences of Calculation and certain chapters of Geometry relative to the calculation of areas. Indeed, it is among the works of the Muslim West that link these two subjects. Its importance hinges equally upon the nature of his materials and mathematical instruments which make it an original book and certainly also significant to this period of transition when three mathematical practices ran in parallel before flowing together into the same stream: that of the East, of Muslim Spain, and of the Maghreb (Djebbar, 1995).

As an example, the following elements contribute to the originality of the work and to its anchorage in the great Arabic mathematical tradition of the 9th-11th Centuries: in Arithmetic, and contrarily to the Maghrebian tradition that would continue from the 14th Century onward, Ibn al-Yasamin treats first multiplication and division before addition and subtraction. In the domain of fractions, he shows, relative to the reading of certain expressions, besides the fraction bar, that the rest of the symbolism was not yet definitively fixed at this epoch. This book is also the oldest that contains both the objects and the operations of Algebra which permit the writing and solution of equations or abstract manipulation of polynomials (Djebbar, 1995).

A fourth major contributor to Maghrebian mathematics was Ahmad Ibn Muncim, who was born in Dénia (on the east coast of Spain, near Valencia) but spent a great deal of his life in Marrakech (Morocco). Ibn Muncim was considered one of the best specialists in Geometry and Number Theory during his time. This mathematician started studying Medicine at the age of 30 and practiced it in Marrakech while teaching and conducting research at the same time. He published many works on diverse subjects that include Euclidian Geometry, calculation, construction of magic squares, Number Theory and Combinatorics. He explored two very important areas of Geometry: (1) the intersections of solids whose bases are curves different from conic sections and (2) curves obtained by the projection of these intersections of solids on a given plane (Djebbar, 1995).

Ibn Muncim also informed readers in the Maghreb from the12th Century onwards about the book of al-Mu'taman (d. 1085) titled *The Book of Perfection*, which is essentially dedicated to Geometry but with the first chapter dealing with Number Theory. This important work was transmitted from the Almohad capital. Later, Maimonides (d. 1204) taught the content of the book of al-Mu'taman in Cairo and also in Fez where he stayed for some time. During the 13th and 14th Centuries, certain chapters of the work continued to be studied in the Extreme Maghreb, particularly by Ibn al-Banna who refers to it explicitly in his *Epistle on the Calculation of Areas* and, a little later, by Ibn Haydar (Djebbar, 1995).

With regard to the content of *The Book of Perfection*, it shows that it does not only concern a simple return to techniques and earlier mathematical results from the Andalusian tradition or transmitted by it. Instead, one finds new trends and results whose origins are to be found in the activities of the Almohad capital or in the preoccupations of its intellectual environment. More precisely, one discovers in it, along with the classical chapters on arithmetical operations, other mathematical operations like the one on the study of figurate numbers, the one on the determination of amicable numbers and, in particular, the one on enumeration of all words of a language utilizing a given alphabet. It is in this work that Ibn Muncim's most important contribution lies, as he dedicates a chapter of 19 pages that contain the important combinatorial propositions and trends which will be rediscovered in Europe only in the 16th and 17th Centuries in particular by Cardano (d. 1576), Mersenne (d. 1648), Frénicle (d. 1675) and Pascal (d. 1662). One may reasonably propose that it was the reinvigoration of Arabic linguistic and grammatical activities in Marrakech that brought the problems of enumeration to the order of the day. It is also reasonable to suggest that Ibn Muncim's way of solving these combinatorial problems by utilizing the concrete model of threads of silk of different colors finds its origins in the industrial or commercial environment of Marrakech (Djebbar, 1995).

Ibn Muncim had students, and one of them, ash-Sharuf, wrote a book titled *The Canon of Calculation* in which he incorporated the combinatorial results of his teacher. Ash-Sharuf, in turn, taught Combinatorics to his pupils, among them was the famous Ibn al-Banna (we will deal a bit more with his own mathematical contributions later). In fact, in Ibn al-Banna's little work titled *Advertisement to the Intelligentsia*, he evokes explicitly one of the methods of Ibn Muncin—that of the arithmetic triangle—to enumerate all words which are possible to pronounce when one utilizes the 28 letters of the Arabic alphabet. But he does not stop here; he introduces three original contributions in this area. The first, which is the most important, concerns the announcement and the demonstration, for the first time, of the formula of factorials giving the combinations of n letters of a given alphabet taking p at a time, without utilizing the arithmetic triangle, a result that will be established once again by Pascal three centuries later. Second, Ibn al-Banna establishes, as far as possible, the relations that exist among the figurative numbers of Nicomachus, the combinations of n objects taking p at a time, and the sums of certain progressions of whole numbers. Third, Ibn al-Banna utilizes the techniques or the trends for combinatorial type to solve certain problems outside mathematics and which lead to enumerations with constraints: for example, the determination of the number of possible readings of a given phrase, taking into account the rules of the Arabic grammar, or of the number of prayers that a Muslim has to say to compensate for the forgetting of a certain number among them (Djebbar, 1995).

In the history of the scientific activities in the Maghreb, the 14th Century marked a privileged moment, both for the quantitative importance of the mathematical production and for the content of this production and influence that it would have, during many centuries, on the teaching of mathematics in the whole of North Africa and sometimes even in certain regions of Sub-Saharan Africa. The majority of the mathematical production of this century is a return—in the form of commentaries, summaries or developments—to a part of what had already been discovered or assimilated during the preceding centuries. New contributions were indeed exceptional. Thus, a fifth major contributor to the Maghrebian mathematical tradition and one of the last innovators of the great Arab mathematical tradition and one of the initiators of a new tradition of teaching mathematics, based on the commentary, is Ibn al-Banna (Djebbar, 1995).

Born in Marrakech in 1256, Ibn al-Banna grew up there and acquired an excellent training in various disciplines. He also lived and taught for some time in Fez which became, after the fall of the Almohads, the capital of the dynasty of the Merinids. Fez tried to rival, on an intellectual level, Marrakech, the only city which had the privilege of having been, for almost two centuries (1062-1248), the capital of the entire Maghreb, including vast Sub-Saharan African zones. A strong scientific tradition was established in Fez (Djebbar, 1995).

Ibn al-Banna was the last of the Maghrebian mathematicians that were involved in research, to the extent that he tackled problems which were new for the epoch, and that he contributed original solutions and advanced new ideas. Already mentioned is his contribution to Combinatorics which lies in the prolongation of research activities and preoccupations of Ibn Muncim. Equally interesting elements are found in his *The Lifting of the Veil* in which he establishes the results that let us understand that during his epoch, the problems of enumeration were not only related to the field of language. He also introduces a new trend in Algebra concerning the justification of the existence of solutions to the canonical equations of al-Khwarizm and he continues with a reflection on bases different from ten that would have been started by Ibn Muncim (Djebbar, 1995).

The importance and prestige of Ibn al-Banna did not only come from his mathematical works. In fact, he distinguished himself from his Maghrebian predecessors by the richness and diversity of his production. Of the more than 100 titles of writings that are attributed to him, only 32 concern mathematics and astronomy. The others deal with disciplines that are very distant from one another, such as Linguistics, Rhetoric, Astrology, Grammar and Logic (Djebbar, 1995).

Among Ibn al-Banna's scientific writings, those related to the Science of Calculation seem to have assured his scientific notoriety. These writings include *The Abridgment*, *The Four Works on Calculation*, and *The Lifting of the Veil*. It is *The Abridgment* that illustrates the trend and conceptions of Ibn al-Banna, at the level of the ordering of the chapters, the conciseness, the rigor, the formulation, and the absence of all symbolism. The factors that led Ibn al-Banna to inaugurate this type of very condensed manual, in comparison to the great collections of the 12th and 13th Centuries, are perhaps not strictly pedagogical. This might have been the consequence of the whole training of Ibn al-Banna and in particular of his mystical guiding that favored a certain esotericism in addition to the difficulty and the abstract character of mathematical notions and techniques (Djebbar, 1995).

The encyclopedic nature of Ibn al-Banna's production may have contributed to his social status, as he was honored by the Merinid power. This led him to leave Marrakech for Fez at the invitation of the sultan of that era. It was this dual status, scientific and social, that helped Ibn al-Banna to solve the problems that preoccupied his contemporaries, and which led him to publish an original book, *Advertisement to the Intelligentsia*, whose contents might be related (because of some of its aspects) to Ethnomathematics. The first part of the book contains the precise mathematical answers to questions that touch varied domains of everyday life such as the composition of medicaments, the calculation of the drop of irrigation canals, the arithmetical explanation of a *sura* (verse) of the Qur'an concerning inheritance, the determination of the hour of the third daily prayer, the explanation of frauds linked to instruments of measurement, the enumeration of delayed prayers which have to be said in a precise order, the exact calculation of legal tax in the case of a delayed payment, etc. The second part of the book, which belongs to the ancient tradition of cultural mathematics, combines a collection of little arithmetical problems presented in the form of poetic riddles. Before Ibn al-Banna, mathematicians had also solved problems related to everyday life; but often, the concrete aspect of a problem was no more than a dressing-up of arithmetical or abstract algebraic exercises. This is not the case for the 17 mathematical problems explored by Ibn al-Banna (Djebbar, 1995).

At the level of the great orientations of mathematical activity during the Middle Ages, Ibn al-Banna appeared as the starting point of a whole tradition that extends to the different regions of North Africa and as far as Egypt, and which remained in Muslim Spain. This tradition is that of commentaries, which have been the object of a comparative study of their contents that reveal both quantitative and qualitative differences. At the quantitative level, one finds short commentaries that explain definitions and algorithms by means of examples without leaving the cadre of the work. At the other extreme, one finds treatises for which the content of the work seems to be just a pretext or a guideline to permit their authors to expose in their own way, while sometimes criticizing Ibn al-Banna severely, the themes dealt with by him and others he had deliberately abandoned. At the qualitative level, these commentaries are distinguishable from one another by the utilization or not of arithmetical and algebraic symbolism and by the recourse or not to the explanation or to the critique of certain definitions, to the demonstration of the propositions evoked by Ibn al-Banna and to the justification of the validity of the algorithms that he exposed (Djebbar, 1995).

The detailed analysis of the most important chapters of these commentaries makes it possible for Djebbar to make further remarks concerning the nature of mathematics taught in the Maghreb during the era. First, the level of the mathematics that is exposed is not lower than that of the previous period, but missing are certain themes that had been taught during the 10th Century, such as the extraction of the approximate cubic root of a number, or the computation of new pairs of amicable numbers. This phenomenon was already perceptible in the work of Ibn al-Banna and it was only to extend from the 14th Century onwards. Second, missing are new contributions in these commentaries, either on the theoretical plane or at the level of the applications of earlier ideas and techniques. The most meaningful novelty is situated at the level of written expression with the progressive use of a relatively elaborate symbolism. Finally, the commentaries of the 14th and 15th Centuries concern the wording and composition of their content and the style that is employed. At this level, two trends are perceptible: (1) the trend characterized by stereotypical formulations which correspond to the usual style of mathematics and (2) the trend which is more rhetorical and is of a higher cultural level in the measure where the authors prolong and complete their mathematical explanations with grammatical, literary or philosophical commentaries. In the domain of Algebra, the works of Ibn al-Banna and other Maghrebian mathematicians continued to be taught in cities such as Fez, Tlemcen, Sebta and Tunis (Djebbar, 1995).

Maghrebian mathematical production of the period that extends from the beginning of the 16th Century to the end of the 19th century has not been well studied. The known number of mathematicians who lived in the Maghreb after the 15th Century surpasses 150. The disciplines they dealt with are Metrical Geometry, Calculation, magical squares, and the distribution of inheritance, as far as mathematics is concerned, and the calculation of time, the determination of the direction of Mecca and the description of astronomical instruments, as far as astronomy is concerned (Djabbar, 1995).

Restricting the discussion to mathematics, the content of the production during this period differs from the earlier mathematical writings in form and standard. The level of the writings is lower than that of the works of the 15th Century, which are themselves less rich, with respect to ideas and techniques, than the works of the 13th and 14th Centuries. This situation corresponds well to the other sectors' intellectual activities in the whole of North Africa. Among the internal factors that were the origins of this long process of decline of mathematical activities are the slowing down and later termination of research and the allure of a qualitative change in the content of the works of teaching that were progressively limited to the exhibition of techniques and results without any demonstration. These internal factors were induced by external ones that were economic in nature, such as the dying out of the African gold routes, the loss of control over sea routes and political spheres, such as the repeated offensives against the coast of North Africa by the new powers of southern Europe: i.e. Spain, Portugal, and certain Italian city-states (Djebbar, 1995).

Therefore, it is not surprising that this long process of degradation had indirect effects on mathematics through the progressive contraction of its domain and of its field of application. Consequently, the only activities that remained for the mathematicians, besides teaching and elaboration of manuals, were those of mathematical practices directly related to activities or preoccupations of a religious nature, such as the distribution of inheritance and donations to rightful claimants, the determination of time for the fixing of moments for prayer, and the construction and use of astronomical instruments such as the quart of sine and the astrolabe (Djebbar, 1995).

The absence of research in mathematics, the contraction of its field of application and the reduction of the content of the programs led to the appearance and development, both in the Maghreb and in Egypt, of the practice of publishing, on a single mathematical subject, a series of works that only differ in style (poetry or prose), volume (book or summary), or form (detailed commentary or glosses). This tendency had already emerged during the second half of the 15th Century in the Extreme Maghreb. It continued at least until the 18th Century. Notwithstanding the absence of originality of these numerous works, however, they constitute precious material for the history of scientific teaching in North Africa (Djabbar, 1995).

CONCLUSION

The preceding review of ten centuries of mathematical activities in the Maghreb makes it possible for Djebbar to provide a few concluding commentaries. First, the mathematical practice in the Maghreb inscribes itself essentially in the Arab tradition. Second, it was essentially the cities of the Extreme Maghreb (in particular Sebta, Fez and Marrakech) which took the relay from Muslim Spain in mathematical activity from the 12th Century onwards until the end of the 14th Century. After this period, there was a greater intervention of two other scientific poles: (1) Tlemcen in the Central Maghreb and (2) Tunis in Ifriqiya. Third, the role of the Maghreb in the diffusion of Arab Mathematics first influenced southern Europe in a direct manner as a result of translations, such as the unique example of the Hebrew translations of the books and in the indirect manner by which the assimilation of local teaching was realized in Arabic followed by the elaboration of manuals or treatises in Latin and Hebrew. The most famous example that illustrates this phenomenon and is still little studied is that of the Italian scholar named Fibonacci (Leonardo of Pisa). As he says himself, he was trained in Bougie (one of the Maghrebian scientific poles of the 12th Century) when he was young and later reproduced in his *Liber Abbaci* certain aspects of the Maghrebian mathematical tradition and in particular with respect to the symbolism and computation of fractions. The transmission of the mathematical writings of the Maghreb essentially went into two other directions. The first direction is that of the East, more precisely Egypt, where the writings were the object of commentaries. The second direction is that of Sub-Saharan Africa. It seems that it was the Extreme Maghreb that was the principal relay for this particular transmission that essentially concerned the Science of Computation and Astronomy. This is confirmed by the manuscripts that are today in the library of Timbuktu (Mali). Among these manuscripts, only one, treating computation, is attributed to a scholar from the region. It concerns a mathematician from Arawan (Mali) who lived after the 16th Century, as he refers in his writing to an arithmetical poem titled "The White Pearl" written by a mathematician of the Central Maghreb. The other manuscripts are either mathematical poems or astronomical writings (Djebbar, 1995).

Chapter Seven

Combinatorics and African Applications

INTRODUCTION

As can be gleaned from this chapter's title, examined here is Combinatorics and how it has been employed to study African mathematical aspects. We begin by providing some background on Combinatorics, including a general definition and brief descriptions of Enumerative, Structural, Extremal, Probabilistic, Geometric and Topological Combinatorics. After that, we discuss some applications of Combinatorics to African mathematical aspects.

The chapter is important because it deals with a mathematical subject whose roots can be traced to Africa and is being resuscitated on the continent, despite many challenges. Thus, it has a strong potential for enhancing the mathematical competence of African students. We will elaborate on the African roots of Combinatorics here and deal with the challenges to resuscitate it on the continent in the section on African applications.

Following N. L Briggs' explanation in his article titled "The Roots of Combinatorics" (1979), the most ancient problem connected with Combinatorics is the house-cat-mice-wheat problem of the Ahmes/Rhind Papyrus (Problem 79). This problem occurs in a similar form in a problem of Fibonacci's *Liber Abaci* and in an English nursery rhyme.

Combinatorics saw its earlier growth in Africa via the Mathematics of Medieval Maghreb. As Ahmed Djebbar (1995) recounts (see more details in the chapter tiled The Maghrebian Tradition), one of the major contributors to Maghrebian Mathematics was Ahmad Ibn Muncim, who was born in Dénia (on the east coast of Spain, near Valencia) but spent most of his life in Marrakech (Morocco). Ibn Muncim was considered one of the best specialists in Geometry and Number Theory during his time. This mathematician started studying Medicine at the age of 30 and practiced it in Marrakech while teaching and conducting research at the same time. He published many works on diverse subjects that include Euclidian Geometry, calculation, construction of magic squares, Number Theory and Combinatorics (Djebbar, 1995).

Ibn Muncim also informed readers in the Maghreb from the12th Century onwards about the book of al-Mu'taman (d. 1085) titled *The Book of Perfection*, which is essentially dedicated to Geometry but with the first chapter dealing with Number Theory. This important work was transmitted from the Almohad capital. Later, Maimonides (d. 1204) taught the content of the book of al-Mu'taman in Cairo and also in Fez where he stayed for some time. During the 13th and 14th Centuries, certain chapters of the work continued to be studied in the Extreme Maghreb, particularly by Ibn al-Banna who refers to it explicitly in his *Epistle on the Calculation of Areas* and, a little later, by Ibn Haydar (Djebbar, 1995).

With regard to the content of *The Book of Perfection*, it shows that it does not only concern a simple return to techniques and earlier mathematical results from the Andalusian tradition or transmitted by it. Instead, one finds new trends and results whose origins are to be found in the activities of the Almohad capital or in the preoccupations of its intellectual environment. More precisely, one discovers in it, along with the classical chapters on arithmetical operations, other mathematical operations like the one on the study of figurate numbers, the one on the determination of amicable numbers and, in particular, the one on enumeration of all words of a language utilizing a given alphabet. It is in this work that Ibn Muncim's most important contribution lies, as he dedicates a chapter of 19 pages that contain the important combinatorial propositions and trends which will be rediscovered in Europe only in the 16th and 17th Centuries in particular by Cardano (d. 1576), Mersenne (d. 1648), Frénicle (d. 1675) and Pascal (d. 1662). One may reasonably propose that it was the reinvigoration of Arabic linguistic and grammatical activities in Marrakech that brought the problems of enumeration to the order of the day. It is also reasonable to suggest that Ibn Muncim's way of solving these combinatorial problems by utilizing the concrete model of threads of silk of different colors finds its origins in the industrial or commercial environment of Marrakech (Djebbar, 1995).

Ibn Muncim had students, and one of them, ash-Sharuf, wrote a book titled *The Canon of Calculation* in which he incorporated the combinatorial results of his teacher. Ash-Sharuf, in turn, taught Combinatorics to his pupils, among them was the famous Ibn al-Banna (I will deal a bit more with his own mathematical contributions later). In fact, in Ibn al-Banna's little work titled *Advertisement to the Intelligentsia*, he evokes explicitly one of the methods of Ibn Muncin—that of the arithmetic triangle—to enumerate all words which are possible to pronounce when one utilizes the 28 letters of the Arabic alphabet. But he does not stop here; he introduces three original contributions in this area. The first, which is the most important, concerns the announcement and the demonstration, for the first time, of the formula of factorials giving the combinations of *n* letters of a given alphabet taking *p* at a time, without utilizing the arithmetic triangle, a result that will be established once again by Pascal three centuries later. Second, Ibn al-Banna established, as far as possible, the relations that exist among the figurative numbers of Nicomachus, the combinations of *n* objects taking *p* at a time, and the sums of certain progressions of whole numbers. Third, Ibn al-Banna utilized the techniques or the trends for combinatorial type to solve certain problems outside mathematics and which lead to enumerations with constraints: for example, the determination of the number of possible readings of a given phrase, taking into account the rules of the Arabic grammar, or of the number of prayers that a Muslim has to say to compensate for the forgetting of a certain number among them (Djebbar, 1995).

COMBINATORICS: SOME BACKGROUND

Combinatorics being a well established sub-area of Mathematics, with its techniques being well used in other academic disciplines, has led to the generation of many studies on the subject. To avoid repetitions in citing works utilized for the discussion in this section, we will simply state that what follows was derived from the following sources which are fully cited in the bibliography: Brualdi (1992), Brusco and Stahl (2005), Constantine (1987), Erdös and Rado (1956), Erdös et al. (1984), Erickson (1996), Godsil (1993), Graham et al. (1990 & 1996), Katz (2007), Krishnamurthy (1986), Kunen (1980), Lawler et al. (1985), Riordan (1958 & 1968), Slomson (1991), Stanley (1997 & 1999), Trotter (1992), Tucker (1984/1995), van Lint and Wilson (2001), and Williamson (1985).

A subfield of Pure Mathematics, Combinatorics (or Combinatorial Analysis, also referred to as the Science of Counting) deals with the selection, arrangement, and operation of mathematical elements—graphs, matrices, lattices, codes, designs, and algorithms—within finite sets. Since Combinatorics deals with concrete problems by limiting itself to finite collections of discrete objects, instead of the more common, continuous Mathematics, no standard algebraic procedures apply to all combinatorial problems; thus, each problem may require a separate logical analysis.

Combinatorial theories deal with existence, enumeration, and construction or structure of elements. These theories include Ramsey's, Pólya's, and the

probabilistic method. For the sake of brevity, not all of these theories are discussed in this chapter. Relevant sources cited above and referenced in the bibliography will serve the interested reader quite well.

Some combinatorial questions involve well-known sets of numbers (e.g., binomial questions) that are on interest in Number Theory: e.g. questions dealing with partitions in Additive Number Theory and questions that concern Finite Fields. Considerable overlap exists between Combinatorics and Group Theory, since graphs and other combinatorial structures have automorphism group. Certain combinatorial problems lead to recursions or generating functions that are treated with, for example, series techniques. Some classical combinatorial questions are geometric in nature: e.g., the theory of polyhedra that provided historical precedent for aspects of Algebraic Topology such as Betti numbers, etc. Fundamental questions in Probability are by their very nature combinatorial. The theory of design is also employed as a topic of experimental design in Statistics. The complexity of combinatorial optimization problems is treated in Computer Science, especially the complexity of Combinatorial Geometry that is studied in Computational Geometry. Combinatorial games are studied in Game Theory. A number of important optimization problems involve choices of a large finite set that is investigated in Operations Research. General principles of Coding Theory are examined in Information Theory. Other fields that have employed Combinatorics include Logic, Linear Algebra, Ordered Structures, and the natural sciences.

Discussions of Combinatorics are generally subsumed into the following six categories: (1) Enumerative Combinatorics, (2) Structural Combinatorics, (3) External Combinatorics, (4) Probabilistic Combinatorics, (5) Geometric Combinatorics, and (6) Topological Combinatorics. The following are brief descriptions of these mathematical aspects.

Enumerative Combinatorics deals with permutations with and without repetitions, combinations with and without repetitions, and Fibonacci numbers. It is used to count the number of ways that certain patterns can be formed. Counting combinations and counting permutations are two examples of this type of problem. In general, given an infinite collection of finite sets $\{S_i\}$ indexed by the natural numbers, enumerative Combinatorics seeks to describe a *counting function* which counts the number of objects in S_n for each n. The simplest such functions are *closed formulae*, expressed as a composition of elementary functions such as factorials, powers, etc. For example, the number of different possible orderings of a deck of n cards is $f(n) = n!$. On many occasions, no closed form is initially available. In such cases, one must first generate a recurrence relation and then solve the recurrence to arrive at the desired closed form. Thus, $f(n)$ may be expressed by a formal power series, called its *generating function*, which is most commonly either the ordinary generating function

$$\sum_{n\geq 1} f(n)x^n$$

or the exponential generating function

$$\sum_{n\geq 1} f(n)\frac{x^n}{n!}.$$

In many cases, a complicated closed formula yields little insight into the behavior of the counting function as the number of counted objects grows. When this is the case, a simple asymptotic approximation may be preferable. A function g(n) is an asymptotic approximation to f(n), if $f(n)/g(n)\to 1$ as $n\to$infinity. In such a case, one writes

$$f(n) \sim g(n)$$

When the generating function is determined, it may allow one to extract all the information given by the previous approaches. Also, the various natural operations on generating functions such as addition, multiplication, differentiation, etc. have a combinatorial significance that allows one to extend results from one combinatorial problem in order to solve others.

Permutations with repetitions are employed when the order matters and an object can be chosen more than once. The number of permutations is

$$n^r$$

where:

n = the number of objects from which you can choose
r = the number to be chosen.

The assumption is that objects have an equal chance of being chosen (1 / n). For example, the number of permutations that can be determined for the letters x, y, and z is 6:

x y z   x z y   y x z   y z x   z x y   z y x

For permutations without repetitions, order matters, and each object can be chosen only once. Therefore, the number of permutations is

$$(n)_r = \frac{n!}{(n-r)!}$$

where:

n = the number of objects from which you can choose
r = the number to be chosen
! = the standard symbol for factorial

For example, if one has four mangoes and is going to choose three out of these, s/he will have $4!/(4-3)! = 12$ permutations.

Note that if $n = r$, meaning the number of chosen elements is equal to the number of elements from which to choose; i.e. four mangoes and pick all four, then the formula is

$$\frac{n!}{(n-n)!} = \frac{n!}{0!} = n!$$

where $0! = 1$.

For example, if one has the same four mangoes and wants to find out how many ways s/he may arrange them, it would be 4! or $4 \times 3 \times 2 \times 1 = 24$ ways. The reason for this is that s/he can choose from 4 for the initial slot and then is left with only 3 from which to choose for the second slot, and so on. Multiplying them together gives the total of 24. Nonetheless, the number of combinations that can be delineated for the x, y and z is one. The order of x y z, x z y, and so on represents the same combination, following the preceding equation.

For combinations without repetitions, when the order does not matter and each object can be chosen only once, the number of combinations is the binomial coefficient:

$$\binom{n}{k} = \frac{n!}{k!(n-k)!}$$

where:

n = the number of objects from which one can choose
k = the number to be chosen

For example, if an African restaurant has ten types of sauces on its menu from which one can chose and the person wants to chose five sauces, there are 10!/5!(10 – 5)! = 252 ways to choose.

In terms of combinations with repetitions, when the order does not matter and an object can be chosen more than once, then the number of combinations is

$$\frac{(n+k-1)!}{k!(n-1)!} = \binom{n+k-1}{k} = \binom{n+k-1}{n-1}$$

where:

n = the number of objects from which one can choose
k = the number to be chosen

For example, if one has 30 students from different grades (n) at a high school and wishes to choose three student leaders (k), s/he would have (30 + 3 – 1)!/3!(30 – 1)!) = 4,060 ways to choose.

The Fibonacci sequence, named after Leonardo Fibonacci, follows the pattern

1, 1, 2, 3, 5, 8, 13, 21, 34, 55, 89 . . . .

where:

$F_{n+2} = F_n + F_{n+1}$ and $n \geq 1$

Starting with $F_1 = 1$ and $F_2 = 1$, each term in the sequence is the sum of the previous: e.g., 34 = 21 + 13.

Structural Combinatorics includes Graph, Design and Matroid Theories. Concerning Graph Theory, graphs are fundamental objects in Combinatorics. The foci of graphs range from counting (e.g., the number of graphs on n vertices with k edges) to structure (e.g., graphs that contain Hamiltonian cycles) to algebraic aspects (e.g., given a graph G and two numbers x and y whether the Tutte polynomial $T_G(x,y)$ have a combinatorial interpretation). While very strong connections exist between graph theory and Combinatorics, the two are on many occasions treated as separate subjects.

In terms of Design Theory, a simple result in the block design area of Combinatorics, for instance, is that the problem of forming sets has a solution only if n has the form $q^2 + q + 1$. It is easier to prove that a solution exists if

q is a prime power. It is conjectured that these are the *only* solutions. It has also been demonstrated that if a solution exists for q congruent to 1 or 2 mod 4, then q is a sum of two square numbers. This last result, referred to as the Bruck-Ryser theorem, is proved by a combination of constructive methods based on finite fields and an application of quadratic forms. When such a structure is nonexistent, it is referred to as a finite projective plane, thereby showing the intersection between finite geometry and Combinatorics.

Matroid Theory abstracts part of Geometry, by studying the properties of sets (usually, finite sets) of vectors in a vector space that do not depend on the particular coefficients in a linear dependence relation. In addition to the structure, but also enumerative properties are also covered in Matroid Theory. For example, given a set of n vectors in Euclidean space, the largest number of planes they can generate is the binomial coefficient,

$$\binom{n}{2}.$$

In essence, in almost all cases, no set exists that generates exactly one fewer plane.

External and Probabilistic Combinatorics include Ramsey Theory for infinite sets, External Combinatorics, Probabilistic Combinatorics, and other topics that are not covered here. Extremal and Probabilistic Combinatorics deal with mathematical aspects concerning set systems. A simple example is to show that if no two subsets are disjoint, the largest number of subsets of an n-element set one can have is half the total number of subsets.

> *Proof*: Call the n-element set S. Between any subset T and its complement $S - T$, at most one can be chosen.

This proves that the maximum number of subsets that can be chosen is not greater than half the number of subsets. To determine how one can attain half the number, s/he can simply choose one element x of S and select all the subsets that contain x.

A challenging problem is to characterize the extremal solutions: i.e. to show that no other choice of subsets can attain the maximum number while satisfying the requirement. Most times, it is difficult even to find the extremal answer $f(n)$ exactly, prompting one to only give an estimate that is asymptotic.

In essence, on the one hand, the type of mathematical aspects addressed in Extremal Combinatorics deals with the largest possible graph which satisfies certain properties. An example is that the largest triangle-free graph on $2n$

vertices is a complete bipartite graph $K_{n,n}$. Probabilistic Combinatorics, on the other hand, deals with the following type of mathematical aspects: the probability of a certain graph property for a random graph (within a certain class); for example, the average number of triangles in a random graph. Probabilistic methods are also employed to determine the existence of combinatorial objects with certain prescribed properties for which explicit examples might be difficult to find, by observing that the probability of randomly selecting an object with those properties is greater than 0.

In Mathematics, Infinitary Combinatorics, also referred to as Combinatorial Set Theory, is an extension of ideas in Combinatorics to infinite sets. Ramsey Theory is one of the approaches used to study such sets. A well-celebrated part of Extremal Combinatorics, Ramsey Theory states that any sufficiently large random configuration will contain some sort of order. Ramsey Theory for infinite sets is written as κ, λ, m and *n* for cardinal numbers. The notation

$$\kappa \rightarrow (\lambda)^n_m$$

is an abbreviation that every partition of the set $[\kappa]^n$ of *n*-element subsets of κ into *m* pieces has a homogeneous set of size λ. It is often omitted when *m* is 2.

Frank Ramsey, for whom Ramsey Theory is named, proved that for every integer *k* there is an integer *n*, such that every graph on *n* vertices either contains a clique or an independent set of size *k*. A frequently cited example is the one about given any group of six people, it is always the case that one can find three people out of this group that either all know each other or all do not know each other. The key given to the proof in this case is the Pigeonhole Principle: i.e. either A knows three of the remaining people, or A does not know three of the remaining people.

Geometric Combinatorics is related to Convex and Discrete Geometry. It investigates the many faces of each dimension a convex polytope can have. Metric properties of polytopes play an important role as well—an example is the Cauchy theorem on rigidity of convex polytopes. Special polytopes also considered include permutoheron, associahedron and Birkhoff polytope.

Topological Combinatorics covers Algebraic Topology, Kneser Conjecture, graph coloring problems, the necklace problem, partially ordered sets, Bruhat orders, Sperner's Lemma, and Discrete Exterior Calculus, among others that are not discussed here for the sake of brevity. In Topological Combinatorics, combinatorial analogs of concepts and methods in topology are employed to investigate graph coloring, fair division, partitions, partially ordered sets, decision trees, necklace problems and discrete Morse theory.

For example, Sperner's Lemma, named after its originator, Emanuel Sperner, is a combinatorial analog of the Brouwer fixed point theorem. The lemma states that every *Sperner coloring* of a triangulation of an n-dimensional simplex contains a cell colored with a complete set of colors. The initial result of this kind was proved in relation with proofs of invariance of domain. Sperner colorings have been used for effective computation of fixed points in root-finding algorithms.

Sperner's Lemma is used to explain one- two- and multidimensional cases. In a one-dimensional case, the lemma is regarded as a discrete version of the Intermediate Value Theorem, which states that if a discrete function takes only the values 0 and 1, begins at the value 0 and ends at the value 1, then it must switch value an odd number of times.

The two-dimensional case, which is the one referred to most frequently, is stated as follows:

> Given a triangle ABC, and a triangulation T of the triangle, the set S of vertices of T is colored with three colors in such a way that
> (1) A, B and C are colored 1, 2 and 3, respectively
> (2) Each vertex on an edge of ABC is to be colored only with one of the two colors of the ends of its edge. For example, each vertex on AC must have a color either 1 or 3.

Then there exists a triangle from T, whose vertices are colored with the three different colors. Generally, there must be an odd number of such triangles.

In the multidimensional case, the lemma refers to a n-dimensional simplex: i.e.

$$A = A_1 A_2 \ldots A_{n+1}.$$

Considered is a triangulation T which is a disjoint division of A into smaller n-dimensional simplices. One denotes the coloring function as $f: S \rightarrow \{1,2,3, \ldots ,n,n+1\}$, where S is again the set of vertices of T. The rules of coloring are as follows:

> (1) The vertices of the large simplex are colored with different colors, i. e. $f(A_i) = i$ for $1 \leq i \leq n+1$.
> (2) Vertices of T located on any given k-dimensional subface

$$A_{i1} A_{i2} \ldots A_{ik}$$

are colored only with the colors

$$f(A_{i1}), f(A_{i2}), \ldots, f(A_{ik}).$$

Then there exists an odd number of simplices from *T*, whose vertices are colored with all $n+1$ colors. There must be at least one simplice.

AFRICAN APPLICATIONS

Perhaps the following E-mail exchange gleaned from the Ubuntu forum (a listserv dedicated to the discussion of scientific and mathematical issues) provides a glimpse into the state of learning and teaching Combinatorics in Africa (http://www.ubuntu-forum.com):

Combinatorics driving me mad

Hello!

I've been sitting here for a while trying to solve/understand a combinatorics example.

I have a set with 12 elements.
I want to create two subsets:
* One with 6 elements
* Another with 7 elements
* 3 elements are shared by both subsets.

In how many ways can this be done?
I've been working with this formulae (sic), but I'm not sure it's the right one:

$$\binom{n}{k} = \frac{n!}{k!(n-k)!}$$

Like first take out 6 elements: 12!/(12-6)!
Then take out 7. The problem is that the set only has 7 elements left. So I'd have to share an element with the first subset.
But then again three elements are to be shared.
This means that there will be elements left in the original set . . .

Would be great if anyone could help me ☺

Thanks!

Re: Combinatorics driving me mad

In case this is homework, I'll just give you a hint. You already know how to count the number of ways you can choose 6 elements from 12. For a given choice, how many ways can you construct the second set? You have to choose 3 from the first set (how many ways can you do that?), and 4 from the 6 that are not in the first set (how many ways can you do that?). Multiply the three numbers together to get the total number of possibilities.

Re: Combinatorics driving me mad

Thanks!

I noticed that I had forgotten the X in: A!/X!(A-B)! which didn't help ☺
It's not homework as such, but I do appreciate that you let me solve it myself. As I want to learn
☺
Thanks a lot ☺ (and I suppose I shouldn't publish my solution then in case anyone else with this problem comes here)

Re: Combinatorics driving me mad

An alternative (equivalent) solution can be obtained by reformulating the problem as

> *In how many ways can I partition a set with 12 elements into four disjoint subsets containing 3, 3, 4 and 2 elements respectively?*

Count the number of ways one can sequentially choose 3 of 12 elements, 3 of 9 elements, 4 of 6 elements and last 2 of 2 elements. Multiply.

Regards, (name deleted).

As can be seen from the preceding exchange, the problem is stated in abstract terms; thus, the answer is provided in a similar manner. A similar tendency can be found in the many attempts to teach and learn Combinatorics in Africa. Even more disappointing is that the many major symposia on Combinatorics that have been organized in Africa neither dealt with the history of the subject in Africa nor explored African mathematical aspects.

For instance, from August 18 to 29, 2003, the West African Summer School in Algebraic Combinatorics convened in the Cape Coast, Ghana dealt with the following topics: the symmetric group and symmetric group presentations, symmetric functions, tableaux, and the computer Algebra packages Maple and GAP (http://garsia.math.yorku.ca). During the East African Summer School on Commutative Algebra and Algebraic Geometry held in Nairobi, Kenya from August 9 to 20, 2004, the topics discussed were division algorithms, Buchberger's algorithm, Groebner bases, and the Algebra computer software CoCoA (http://garsia.math.yorku.ca). At the Third Eastern African Workshop on Algebra, Geometry and Combinatorics held in Kampala, Uganda from June 19 to 30, 2006, the focus was on commutative Algebra and Combinatorics (http://www.mat.unicoma1.it). And during the 6th Symposium on Training and Research on Pan African Mathematics Olympiads convened in Ouagadougou, Burkina Faso from November 5 to 10, 2007, the workshops dealt with Algebra and Number Theory, Combinatorics, Geometry, and Inequalities, employing Western Toolkits (PAMOSTAR, 2007). It is interesting to note that all of these symposia were sponsored by Western institutions and most of the facilitators were also from the West.

Nonetheless, a number of scholars have employed Combinatorics to explore African mathematical aspects, mostly those dealing with African board games. Some of these are discussed in the paragraphs that follow. Since the authors of these works use different approaches to present their findings, they are discussed separately for the sake of coherence. The works of a number of other scholars are not easily accessible; thus, these works are mentioned but not discussed. It is worth noting that many of the board games were brought by Africans to and can still be found in the African Diaspora. Examples of these include the *Barbados Warri* and the Yoruba *Ayoayo* which came to be known in Barbados as *Round-and-Round Warri* studied by W. Lee Farum-Badley (n.d.) and the Surinam *adji boto* mentioned by Claudia Zaslavsky (1973/1999:118).

Claudia Zaslavsky, while not explicitly stating the word Combinatorics in describing the Mathematics of the games she examines, employs techniques that are common in Combinatorics in her discussion of counting rhymes and rhythms, tic-tac-toe or three-in-a-row, networks, riddles, arrangements, games of chance, two-row and four-row board games, and magic squares. Each of these mathematical aspects is briefly discussed for the sake of brevity (Zaslavsky, 1973/1999:102–152).

For counting rhymes and rhythms, Zaslavsky recounts that African children learn finger counting rhymes even before becoming aware of the number sequence. By providing many examples from around Africa, she shows that some rhymes go to five, others continue up to 20, but most go as far as ten to correspond to the number of fingers. In some communities, the rhymes are based on a 12-system or emphasize multiples of three or four (Zaslavsky, 1973/1999:102).

In terms of tic-tac-toe or three-in-a-row, Zaslavsky notes that several versions of the game (some older than 3,300 years) can be found throughout Africa and all of them are more complicated than the familiar "noughts and crosses." The game is played on either a board or on lines drawn in the ground and proceeds in three stages. In the first stage, each player in turn places a stone at an unoccupied intersection. The goal is to form a line of three counters in any direction. The player who is able to form a line of three can remove any one of the opponent's stones. Once a player has formed a line of three, s/he could, on the subsequent move, remove one of the stones in the line and place it at a different intersection on the next move in order to capture another of the opponent's counters. The second stage begins when each player has placed all twelve of his/her stones on the board. Both players then proceed to move the stones one space at a time along a line; the goal again is to complete lines of three. It is obvious that some of the stones would have been captured; otherwise, the 24 stones would be locked into position at the 24 intersection, thereby preventing any movement. The final stage occurs when one of the players has three stones left. S/he can then move a stone to any free intersection point on the board. When an opponent is left with two stones, s/he loses the game (Zaslavsky, 1973/1999:103–104).

As stated earlier, there are several versions of the tic-tac-toe or three-in-a-row game. One version is the "African Morris" played by the Asante, with rules similar to those played in Zimbabwe. A 3,300-year old version was found on a board that was cut into a roof slab of the ancient Egyptian temple at Kurna. A similar game called The Mill or Nine Men's Morris goes beyond someone being the first to make three-in-a-row, as the game of placement is combined with that of movement. Another Zimbabwean version named *Tsoro Yematatu* has seven intersections, with each player using just three stones that can be moved anywhere. The first to complete a line of three wins the game. It is much more difficult, as there is only one free intersection available once all of the stones have been placed (Zaslavsky, 1973/1999:104–105).

As it pertains to networks, Zaslavsky retells the story of the renowned Hungarian adventurer Emil Torday who was unable to draw a group of figures in the sand without lifting his finger or retracing any line segment, as requested by children in the Congo who could do it with much ease. According to Zaslavsky, to determine whether a figure can be retraced, one must examine the vertices to see how many are odd and how many are even. A network with exactly two odd vertices can be retraced by a single path: i.e. drawn without lifting the finger or retracing. One of the odd vertices can serve as the starting point and the other as the stopping point. A network with no odd vertices can be traversed by a single path no mater where one starts (Zaslavsky, 1973/1999:705–706).

For riddles, Zaslavsky uses an example of a familiar story told by the Kpele children of Liberia. It is about a man who has a leopard, a goat, and a bunch of cassava leaves to be transported across a river, with the stipulation that the boat can carry no more than one of the items at a time including the man. Left with the leopard, the goat will be eaten; left with the goat, the cassava leaves will be imbibed. Since the leopard does not eat cassava leaves, they are the only pair that can be left alone together. Thus, the man must first ferry the goat across the river. Zaslavsky then provides a diagram of one possible solution (1973/1999:109-110), which can be stated as follows: first, the man takes the goat across the river and returns alone to the leopard and cassava leaves; second, he takes the cassava leaves across the river, leaves them there, and returns with the goat; third, he leaves the goat at the starting point and takes the leopard across the river; finally, he returns to the starting point, takes the goat, and crosses the river to join the leopard and the cassava leaves.

In terms of arrangements, Zaslavsky recounts a game Kpele children play that involves lining up 16 stones in two rows of eight each. One child is sent away and the others choose a stone. When the child returns, s/he must determine which stone was selected. S/he may ask four times in which of the two rows the stone is located. After each response, s/he may rearrange the stones within the two rows and must be able to identify the chosen stone after the fourth response (Zaslavsky, 1973/199:110).

For games of chances, Zaslavsky provides several examples of gambling games played by men in most of Africa. These games are based on the outcomes when chips, nuts, or cowrie shells are tossed, with procedures frequently resembling those of a diviner. They are similar to the games of dice and coin-tossing. Unlike the symmetry of a coin in which head or tail is equally likely to occur, a similar prediction cannot be made for the asymmetrical cowrie. Further research would be required before giving odds, a probability calculation many African men have gotten down to a science (Zaslavsky, 1973/1999:113).

As pertaining to two-row and four-row versions of the board game that has come to be generally named *Mancala*, an Old Egyptian word meaning "transferring," Zaslavsky notes that there are hundreds of names and dozens of versions of it throughout Africa and other parts of the world. It is played by kings on beautifully carved ivory boards decorated with gold and by children in holes scooped out of the earth, using pebbles or seeds as counters. By programming electronic digital computers, engineers are learning how machines can make decisions. If a *Mancala* is played with just 36 counters distributed in two rows of six holes, there are about 10^{24} (a million times a million times a million times a million) total possibilities (Zaslavsky, 1973/1999:118).

Finally, in terms of magic squares, Zaslavsky provides many examples to show that the construction of such squares is a mathematical recreation based on Number Theory and that their magic lies in the mystifying arrangement of the numbers so that the sum of each row, each column, and each of the two diagonals is constant. One of the major pioneers of magic squares was

Muhammad ibn Muhammad, an early 18th Century astronomer, mathematician, mystic and astrologer living in Katsina (now northern Nigeria). Anyone interested in his work can find details in Zaslavsky's book and other sources she cites pertaining to him (Zaslavsky, 1973/1999:136–151).

In his article titled "Omweso: The Royal Mancala Game of Uganda—A General Overview of Current Research" (n.d.), Brian Wernham summarizes existing findings (of particular interest in the current study are those by Mayega, 1974; Ilukor, 1978; and Donkers et al., 2001) and offers new ones on the Mathematics of the *Omweso* board game. According to Wernham, *Omweso* is a popular *Mancala* game played in Uganda, with major tournaments held in the capital city of Kampala. The game differs from similar ones played in West Africa because it is a "re-entrant" variety: i.e. all of the seeds remain in play, as captured seeds are re-entered onto the winner's side of the board. Also, unlike the similar game of *Bao* played on the Swahili coast of East Africa, in *Omweso,* players start with all 64 seeds in play and set up freely at the start (Wernham, n.d.:1–2).

The objective of *Omweso* is to capture an opponent's seeds until s/he is unable to move or to gain a knockout by "Cutting off both his heads": i.e. to capture seeds from both ends of his/her board in a single move. In earlier times, games were slower, with players thinking for many minutes to make the right moves. In modern times, fast play is required, with only three seconds thinking time per turn. The referee counts "Omu," "Eberi" and if a player does not choose a hole to start his move, s/he loses his/her turn to the opponent (Wernham, n.d.:2–3).

The following are the four rules that govern the play of *Omweso* (Wernham, n.d.:3–4):

Rule 1—sowing seeds: choose a hole to sow from on your side of the board; pick up all the seeds therein; sow them on anti-clockwise; if the last ends in an empty hole end of your turn; singletons cannot move

Rule 2—relaying sowing: if the last is not sown into an empty hole, then . . . ; pick up all the seeds in that hole . . . ; and sow again!; (and again and again until an empty hole is found . . .)

Rule 3—capture: pairs of occupied holes are vulnerable; lower's inside hole opposite is occupied—this acts as a "marker" for the attack; upper's six seeds are captured . . . ; . . . and re-enter the winner's side of the board AS IF FROM THE ORIGINATING HOLE

Rule 4—reverse capture: reverse capture is allowed from leftmost four holes; clockwise movement from these special holes—BUT ONLY IF CAPTURE IS POSSIBLE

The following are the two ways to win in the street play of *Omweso* (Wernham, n.d.:3–4):

(1) Normal—the loser is immobilized and cannot move (only has empty holes and singleton seeds)
(2) *Emitwe-Ebiri*—"cutting off at 2 heads." Capture of both extreme pairs of holes in one move

There are also special tournament rules (Wernham, n.d.:4):

(1) *Okukoneeza*: if a hole has 3 seeds during the opening before first capture two seeds into next hole, final seed into next but one
(2) Winning by *Akakyala*—capturing seeds from the loser in two separate moves before the loser has even made his first capture of the game

Wernham also notes that another "strong" variant of *Omweso* is the Rwandese game of *Igisoro*, which has reversing rules that make the game quite complex (n.d.:6).

Wernham discusses three types of complexity in *Omweso*. The first is *state-space complexity*, with the following formula (n.d.:8):

$$[(h - 1 + s)!]/[(h\text{-}1)! \times s] \qquad \text{minus } k$$

where:

h = number of holes
s = number of seeds
k = number of illegal/improbable positions

Each player has 32 seeds to set up, giving 7.5×10^{11} possible positions for the first player to set up, which can be countered by the opponent in 7.5×10^{11} ways, giving 5.6×10^{23} combinations. There are no illegal set-up combinations and very few improbable ones in tournament play (Wernham, n.d.:8).

The second is *game-tree complexity*, which can be calculated as follows (Wernham, n.d.:9):

$$i_1 \times i_2 \times (b)^p$$

where:

i = branches in set-up of game for players 1 and 2
b = branches per move
p = plays in game length

During the 2000 Kampala tournament, the average in a game was 5.4 anti-clockwise possibilities plus the possibility of deciding to reverse capture in about 20% of sowings. Thus, Wernham calculates the number of branches per player turn in *Omweso* to be well above six. Assuming that there are seven branches per turn with an average game being 60 turns, the branching complexity would become $7^{60} = 5 \times 10^{50}$ (Wernham, n.d.:9).

The third is *mutational complexity*, referring to the number of changes on the board due to a single move. According to Wernham, the average number of changed holes in the sample game was exactly three per move for the first nine moves while the players positioned themselves without laying themselves open. The next stage had an average of 7.1 moves per player turn, as one player maintained a large position ready to strike, and thereafter 13.4 per turn. The average number of holes changed across the sample *Omweso* game was 9.31, 12.4 if one double-counts holes sowed into more than once a turn. The game was 26 moves per player and 3.05 minutes in length (Wernham, n.d.:10).

Next, Wernham discusses the standard set-ups of *Omweso* using two examples, acknowledging that there are many standard openings in the game. One example he calls the "junior grouping" set-up, whereby eight seeds or less are used, and the other example he dubs the "senior grouping" set-up, whereby 23 seeds are utilized. He notes that there is no restriction on setting up seeds, albeit in street play one generally only uses junior groupings. Junior groupings having a maximum of 5, 6, or 7 seeds are popular, and senior groupings of 17, 19, 20, 21, 22 and 23 seeds are frequently used (Wernham, n.d.:11-12).

Following the discussion on standard set-ups is that on magic numbers. Here, Wernham first draws from the work of J. V. Mayega who wrote the first known *Omweso* computer program using Algol 60 and presented some statistical analysis in his 1974 paper with some work on applying matrix theory to the *Omweso* problem to create a goal-seeking program driven by a points-scoring system. Wernham next draws from the work of Y. Ilukor who took a different approach by investigating the magic numbers within *Omweso* instead of computing traditional statistical and matrix bound techniques (both of these scholars' works are cited in the bibliography). The following is a summary of the generated magic squares (Wernham, n.d.:13–14):

Magic number zero:
The difference of inner outer diagonal sums
$(15 + 6) - (14 + 7) =$ zero

Magic number 2:
(a) The differences of one player's diagonal sums
$(15 + 3) - (14 + 2) = 2$

(b) The differences of inner diagonal sums
$(5 + 5) - (4 + 4) = 2$
(c) The differences of outer diagonal sums
$(15 + 11) - (14 + 10) = 2$

Magic number 8:
(a) The difference of opposite inner/outer row holes
$15 - 7 = 8$
(b) The differences of diagonal inner/outer multiples
$(10 \times 3) - (11 \times 2) = 8$

Magic number 9:
(a) The sum of inner opposites
$3 + 6 = 9$
(b) The differences of inner diagonal multiples
$(7 \times 3) - (6 \times 2) = 9$

Magic number 17:
(a) The sum of opposite outer row holes
$14 + 3 = 17$
(b) The differences of outer hole diagonal multiples
$(2 \times 16) - (1 \times 15) = 17$

Magic number 25:
(a) The sum of outer opposites
$15 + 10 = 25$
(b) The differences of outer diagonal multiples
$(12 \times 14) - (11 \times 13) = 25$

Finally, Wernham presents trivial and complex never-ending moves with attendant proofs for *Omweso*. According to him, trivial settings have the following properties (Wernham, n.d.:15):

(a) One can see the repetition immediately without experimentation
(b) After each move the board is left in exactly the same state, except that the starting hole position is rotated
(c) Triangular numbers play a part with patterns of 4, 3, 2, 1 appearing

Complex positions have the following attributes (Wernham, n.d.:16):

(a) A large (200+) number of repetitions before the position appears again
(b) When the seeds appear again in the same sequence, the starting hole is in exactly the same position (i.e. no intervening rotation occurs)

(c) The number of iterations in these examples is always divisible by 4
(d) There are no obvious patterns occurring

Wernham concludes by raising many social and mathematical questions and stating that *Omweso* poses a unique and interesting area of board games research (n.d.:19).

Johnson Ihyeh Agbinya in his book, *Computer Board Games of Africa: Algorithms, Strategies and Rules* (2004), attempts to provide a general treatise on board games across the African continent by employing techniques from the fields of Mathematics and Computer Science. Since the preceding work by Wernham covers a description of a board game (moves, capturing of seeds, and winning), algorithmic representations or game complexities, magic numbers, trivial and complex never-ending moves, and junior and senior defining matrices, they are not discussed here vis-à-vis Agbinya's work, since both scholars drew from the same material for their works. Likewise, since the preceding work by Zaslavsky discusses additive series and discrete self-organization, they are also not discussed here. Instead, the discussion that follows is limited to Agbinya's analyses of the fundamentals of African board games and an aspect of the Mathematics of the games that Zaslavsky and Wernham do not address—specifically, symmetrical series.

According to Agbinya, African board games have distinct names (he provides a short list of 40 names), rules, strategies and styles that vary from region to region. He finds it unfortunate that some non-African writers have lumped together all versions of count and capture games under the generic name of *Mancala*. He also notes that some authors call them *Manqala, Mankala* or *Mankaleh* which he says are hardly used by Africans. He adds that *Mankaleh* is a derivative of Swahili, a hybrid language that has a great deal of Arabic borrowings (Agbinya, 2004:17).

For Agbinya, since African board games are games of strategy, full information, logic and intelligence, it is imperative to ask questions of intelligence, logic and mathematical reasoning when investigating them. Thus, he asserts that they are games of intellect and thinking rather than chance. Their key strengths, according to him, are (a) strategic thinking, (b) employment of mathematical skills, (c) employment of logical skills, (d) ability to plan and forward thinking in playing, and (e) ability for abstract thinking (Agbinya, 2004:19).

All African board games, according to Agbinya, involve shift, add and subtract operations. Usually many of the moves, he suggests, can be represented with a function in which the numerical values increase by one at a time except one value that is decreased to zero entirely (Agbinya, 2004:31). He defines a move as "a discrete expression in which the number of holes reached by sowing new seeds is given by n and $x(k)$ denote the number of seeds in hole

k at the beginning of iteration 1 and $y(k)$ the number of seeds in hole k at the beginning of iteration 2" (Agbinya, 2004:27). Thus:

$$Y(k) = x\ (k) + n$$

The content of a hole increases by 1 due to sowing if the integral value n is large enough to reach the hole after deposing seeds in its predecessor holes.

Agbina discusses three types of symmetrical series that arise from the *Omweso* game. These series, according to him, have the following two characteristics (Agbinya, 2004:37):

(a) In each case a pair of identical series is observed.
(b) The symmetry of the series is across the middle vertical dividing line. This line divides the board into two 4 x 4 matrices as:

$$O(4,8) = L(4,4){:}\ R(4,4)$$

where $L(4,4)$ is the left 4 x 4 and $R(4,4)$ is the right 4 x 4 matrices of the *Omweso* board.

The first type of these series is the *symmetrical odd integer square series* that is formed across the board by multiplying differences of holes in columns. The difference of outer holes in each of the columns multiplied by the difference of inner holes in the same column is a square sequence of the odd numbers between 1 and 7. The second type is the *decreasing symmetrical integer series* that is obtained by subtracting the sum of inner holes from the outer holes in each column. The third type is the *symmetrical mirror image even series* which emerges when products of the inner holes for each column are delineated (Agbinya, 2004:37-38).

Finally, in their article, "The Combinatorics of Mancala-Type Games: Ayo, Tchoukaillon and 1/π" (2008), Duane Broline and Daniel Loeb examine the two-player game *Ayoayo,* or *Ayo* for short, played by the Yoruba of western Nigeria and the solitaire game *Tchoukaillon* (a Russian game of possible Paleosiberian or Eskimo origin) as a variant of the game *Tchouka* played in central Europe. The authors analyze certain common unbalanced *Ayo* endgame positions which they call *"determined"* and demonstrate how they are related to the positions in *Tchoukaillon* from which a win is possible. They postulate that for all s, there is a unique such position with s stones. They note that certain positions are not realizable on a finite board with a fixed number of pits. They show, however, that the number of stones in such a position on a board with $2n$ pits is bounded by approximately n^2/ϖ. They further study the actual distribution of stones into pits and discover a periodicity in the contents of the first k pits (with respect to the total number of stones) of 1cm—1, 2, . . . , $k = 2$ (Broline and Loeb, 2004:0-1). Here, I focus only on the *Ayo* game since this study is about African Mathematics.

Ayo is played on a wooden board that is approximately 20 inches long, eight inches wide, and two inches thick. Carved on the board are two rows of six pits each about three inches in diameter. Either dried palm nuts or, more commonly, the stones of the shrub *caesalpina crista* are used as playing pieces. The following are the rules of the game (Broline and Loeb, 2004:1-2):

Set up: 48 stones are used. Initially, 4 are placed in each of the 12 pits. (Broline and Loeb generalize the game somewhat allowing boards with $2n$ pits and arbitrary placements of stones.)

Players: Two players alternate making moves. Each player's side of the board has n pits.

Objective: The object of the game is to capture the most stones.

Movement: To move, a player chooses a non-empty pit from his or her side of the board, and removes all of its stones. The stones are redistributed (*sown*), one per pit, among the pits in a counterclockwise direction beginning with the pit after the chosen pit.

Odu: A pit which contains $2n$ or more stones is said to be an *Odu*. If the chosen pit is the *Odu*, the redistribution proceeds as usual except that the initial pit is skipped on each circuit of the board. (Broline and Loeb consider none o f the positions to contain an *Odu*.)

Capture: If the last pit sown by a player is on the opponent's side of the board and contains (after having been sown) two or three stones, then the stones in this pit are captured. Also captured are stones in the consecutively preceding pits which meet these conditions.

End of Game: At each turn, a player must, if possible, move in such a way that his or her opponent has a legal move. If, on some move, a player cannot move in such a way to give his or her opponent a legal move, the game is over and the player is awarded all remaining stones. If there are so few stones on the board that neither player can ever capture, but both players will always have a legal move, the game is over and each player is awarded the stones on his or her own side of the board. For example, if the position is no further captures are possible, but each player can always move to give the opponent a legal move. In this case, each player is awarded a single stone.

Broline and Loeb note that the game begins rapidly with both players demonstrating their dexterity and skill by the speed of their movements. To play the game well, however, a player must remember the number of stones in each of the 12 pits in addition to planning several moves in advance. All this makes the opening game interesting to watch but difficult to learn (Broline and Loeb, 2004:2).

The authors also observe that the endgame is less exciting but easier to analyze, as the latter stages of the game tend to be dominated by one player. They note that the player can move in such a way that his/her opponent has at all times only a single legal move; and that after this sequence of moves, only a few stones are usually left and no further captures can occur (Broline and Loeb, 2004:2).

As noted earlier, Broline and Loeb analyze a specific type of endgame on a generalized *Ayo* board with $2n$ pits numbered clockwise as follows: $-n + 2$, $-n + 1, \ldots, -1, 0, 1, \ldots, n, n + 1$. The authors denote the two players as S for South and N for North. They postulate that S makes his/her plays from pits numbered from $n + 1$ down to 2, while N makes his/her plays from pits numbered from 1 down to $-n + 2$. Play is proposed to proceed from higher numbered ones and from pit $-n + 2$ to pit $n + 1$. Broline and Loeb then provide the following definition for the endgame positions (2004:2–3):

Definition 1 A determined position *is an arrangement of stones on a generalized Ayo board where it is possible for S to move such that*

(a) S captures at every turn,
(b) there is no move from an Odu,
(c) after every turn, N has only one stone on his side of the board, and
(d) all stones are captured by S except one which is awarded to N.

Asserting that it is a simple matter to establish the contents of the pits on N's side of the game board in a determined position, Broline and Loeb sugest the following lemma and attendant proof (2004:3):

> *Lemma 2 The stone on N's side of a determined Ayo position must be in pit 1 if N is to move, and in pit 0 if S is to move.*
>
> *Proof:* If S is to move, she must capture and leave only one stone. Thus, the stone captured must lie in N's second pit (pit 0). Hence, before N's move, the stone must have been in pit 1.

While Broline and Loeb limit their study to the determined positions in the *Ayo* game, they do lay the foundation for one to investigate strategies designed to maximize the number of stones captured in a game.

CONCLUSION

The preceding discussion clearly shows that a student studying African mathematical aspects (which are all around Africans) using Combinatorics will help him/her to develop powerful problem solving and theory building skills. Most of the aspects investigated in Combinatorics have numerous natural connections to other areas. For example, a student studying Computer Science will benefit from Combinatorics because it is used in that field to obtain estimates on the number of elements of certain sets.

Chapter Eight

Vector Calculus and African Applications

INTRODUCTION

As its title suggests, examined in this chapter is Vector Calculus and how it has been employed to study African designs. We begin by providing some background on Vector Calculus, including a general definition, operations, theorems, identities, and fields. After that, we discuss the applications of Vector Calculus to African designs.

The importance of this chapter hinges on the fact that many quantities which are of interest in the study of physical African designs are directed quantities (vectors) and can take on a continuous range of values, making calculus methods imperative in investigating them. Several questions from Vector Calculus become particularly important in understanding those physical designs and solving their concomitant problems.

VECTOR CALCULUS: SOME BACKGROUND

Vector Calculus is a well established subfield in Mathematics and its techniques are well used in other sciences. This has led to the production of many works on the subject. To avoid redundancy in citing works employed for the discussion in this section, I will simply state that what follows was derived from the following sources which are fully cited in the bibliography: Acheson (2005), Arfken and Weber (2005), Corwin and Szczarba (1995), Crowe (1994), Griffiths (1999), Marsden and Tromba (2003), Price (1984), Schey (2005), Tai (1995 & 1997), and Young (1993).

Vector Calculus, also known as Vector Analysis, is generally defined as a branch of Mathematics that investigates multivariable real analysis of vectors in an inner product space of two or more dimensions, albeit some results that pertain to the cross product are only applicable to three dimensions. With its origin in Quantum Analysis, Vector Calculus has developed many formulae and problem solving techniques that are of import to the fields of Engineering and Physics.

Since the concept *vector* itself has several meanings, it makes sense to define it further to distinguish its meaning in Vector Calculus from its meanings in other pursuits. Due to the fact that *scalars* are quantities that only have a magnitude like mass, speed, and electric field strength, it becomes imperative to have a quantity that has not only a magnitude but also a direction; this type of quantity is called a *vector*. Quantities represented by vectors include velocity, acceleration, and virtually any type of force (frictional, gravitational, electric, magnetic, etc.). It should be noted that all of these quantities not only have a magnitude (such as speed: the magnitude of the velocity vector), they also occur or act in a given direction.

An example of when vectors are necessary is the following: Suppose a helicopter traveling at 250 miles per hour to the south with no wind present encounters an easterly crosswind of 45 miles per hour. The resultant velocity of the helicopter would be the sum of the velocities of the wind and the plane. To find this resultant velocity, vectors must be utilized.

Vectors are presented with ordered pairs in pointed brackets to distinguish them from ordered pairs in normal parentheses which represent points. The vector <1,4> is a two-dimensional vector, or directed line segment, from any point (x,y) to the point (x+1,y+4). Likewise, the vector <a,b,c> is a three-dimensional vector from any point (x,y,z) to the point (x+a,y+b,z+c). It is imperative to bear in mind that a vector is independent of its position in the coordinate system.

The magnitude (or length) of a vector v with initial point (x_1,y_1,z_1) and terminal point (x_2,y_2,z_2) is as follows:

$$|\vec{v}| = \sqrt{(x_2 - x_1)^2 + (y_2 - y_1)^2 + (z_2 - z_1)^2}$$

Vectors obey the natural intuitive laws of addition and scalar multiplication:

$$c < x_1, y_1 > + d < x_2, y_2 > = < cx_1 + dx_2, cy_1 + dy_2 >$$

There are two ways to multiply two vectors and get a useful result. One approach produces a scalar (the *dot product*), and the other approach produces a vector (the *cross product*). The *dot product* of two vectors a=<a_1,a_2,a_3> and b=<b_1,b_2,b_3> is derived by the following:

$$\vec{a} \cdot \vec{b} = a_1 b_1 + a_2 b_2 + a_3 b_3 \qquad (*)$$

Another definition of the *dot product* is this:

$$\vec{a} \cdot \vec{b} = |\vec{a}| \left|\vec{b}\right| \cos\theta \qquad (**)$$

where θ (theta) is the angle between the two vectors and |c| represents the magnitude of the vector c.

Mathematically, the scalar projection of *b* onto *a* is |*b*|cos( ) (where is the angle between *a* and *b*) which form (*). This quantity is also called the component of *b* in the *a* direction and, thus, the vector projection is merely the unit vector *a*/|*a*| times the scalar projection of *b* onto *a*. Therefore, the scalar projection of *b* onto *a* equals the magnitude of the vector projection of *b* onto *a*. The second definition is useful for finding the angle between two vectors. An important use of the *dot product* is to test whether or not two vectors are orthogonal: i.e. if the angle between them is 90 degrees. Using (**), the *dot product* of two orthogonal vectors is zero. In essence, two non-zero vectors have dot product zero if and only if they are orthogonal.

The *cross product* is a vector product that yields another vector rather than a scalar. As with the *dot product*, the *cross product* of two vectors contains valuable information about the two vectors themselves. The *cross product* of two vectors *a*=<a_1,a_2,a_3> and *b*=<b_1,b_2,b_3> is delineated by the following:

$$\vec{a} \times \vec{b} = < a_2 b_3 - a_3 b_2, a_3 b_1 - a_1 b_3, a_1 b_2 - a_2 b_1 >$$

While this definition may appear banal, its useful properties become evident when one recalls that the determinant of a 2 × 2 matrix is as follows:

$$\begin{vmatrix} a & b \\ c & d \end{vmatrix} = ad - bc$$

and the determinant of a 3x3 matrix is the following:

$$\begin{vmatrix} c_1 & c_2 & c_3 \\ a_1 & a_2 & a_3 \\ b_1 & b_2 & b_3 \end{vmatrix} = c_1 \begin{vmatrix} a_2 & a_3 \\ b_2 & b_3 \end{vmatrix} - c_2 \begin{vmatrix} a_1 & a_3 \\ b_1 & b_3 \end{vmatrix} + c_3 \begin{vmatrix} a_1 & a_2 \\ b_1 & b_2 \end{vmatrix}$$

One can then write the formula for the *cross product* as follows:

$$\vec{a} \times \vec{b} = \begin{vmatrix} a_2 & a_3 \\ b_2 & b_3 \end{vmatrix} \hat{i} - \begin{vmatrix} a_1 & a_3 \\ b_1 & b_3 \end{vmatrix} \hat{j} + \begin{vmatrix} a_1 & a_2 \\ b_1 & b_2 \end{vmatrix} \hat{k} = \begin{vmatrix} \hat{i} & \hat{j} & \hat{k} \\ a_1 & a_2 & a_3 \\ b_1 & b_2 & b_3 \end{vmatrix}$$

In essence, Vector Calculus is employed to study scalar fields, which associate a scalar to every point in space, and vector fields, which associate a vector to every point in space. For example, the temperature of a lake is a scalar field: to each point, one can associate a scalar value of temperature. The water flow in the same lake is a vector field: to each point, one can associate a velocity vector.

Vector Calculus examines various differential operators defined on scalar or vector fields that are typically expressed in terms of the Del Operator, characterized as the collection of partial derivative operators and calculated as follows:

$$\nabla = \frac{\partial}{\partial x} i + \frac{\partial}{\partial y} j + \frac{\partial}{\partial z} k$$

The following are brief descriptions, domain, range and mathematical notations of the four most commonly used operations in Vector Calculus.

(1) *Curl* measures the tendency of a rotation about a point in a vector field and maps vector fields to vector fields:

$$\nabla \times E = \left(\frac{\partial E_z}{\partial y} - \frac{\partial E_y}{\partial z} \right) i + \left(\frac{\partial E_x}{\partial z} - \frac{\partial E_z}{\partial x} \right) j + \left(\frac{\partial E_y}{\partial x} - \frac{\partial E_x}{\partial y} \right) k$$

(2) *Divergence* measures the magnitude of a source or sink at a particular point in a vector field:

$$\nabla \cdot E = \frac{\partial E_x}{\partial x} + \frac{\partial E_y}{\partial y} + \frac{\partial E_z}{\partial z}$$

(3) *Gradient* measures the rate and direction of change in a scalar field and maps scalar fields to vector fields:

$$\nabla f = \frac{\partial f}{\partial x} i + \frac{\partial f}{\partial y} j + \frac{\partial f}{\partial z} k$$

(4) *Laplacian* is a composition of the divergence and gradient operations and maps fields to scalar fields:

$$\nabla \cdot \nabla f = \nabla^2 f$$

It is imperative that both a magnitude and a direction be specified for a vector quantity, as opposed to a scalar quantity which can be quantified with just a number. Basic vector operations can combine any number of vector quantities of the same type: i.e. same units. When both the domain and range of a function are multivariable, such as a change of variables during integration, the Jacobian quantity is useful.

Similarly, there are four most commonly referenced Vector Calculus theorems. The following are brief descriptions and statements of these theorems.

(1) *Divergence Theorem* postulates that the integral of the divergence of a vector field over some solid equals the integral of the flux through the surface that bounds the solid:

$$\oint \nabla \cdot E dV = \oint E \cdot dA$$

(2) *Gradient Theorem,* also known as the *Fundamental Theorem of Calculus for Line Integrals,* states that the line through a gradient, or vector, field equals the difference in its scalar field at the endpoints of a curve:

$$\phi(\mathbf{q}) - \phi(\mathbf{p}) = \int_L \nabla\phi \cdot d\mathbf{r}$$

(3) *Green's Theorem* suggests that the integral of a scalar curl of a vector field over some region in a plane equals the line integral of the vector field over the curve that bounds the region:

$$\int_C L\,dx + M\,dy = \iint_D \left(\frac{\partial M}{\partial x} - \frac{\partial L}{\partial y}\right) dA$$

(4) *Stokes' Theorem* states that the integral of a curl of a vector field over a surface equals the line integral of the vector field over the curve that bounds the surface:

$$\oint \nabla \times E \cdot dA = \oint E \cdot dL$$

In lieu of these theorems, Vector Calculus calls for the handedness of the coordinate system to be taken into account. The interested researcher can consult the works of Schey (1997) and Griffiths (1999) cited in the references that thoroughly discuss cross product and handedness for details.

Identities are also vital in Vector Calculus. The following are the ones most used by researchers.

(a) The divergence of the curl is equal to zero:

$$\nabla \cdot (\nabla \times B) = 0$$

(b) The curl of the gradient is equal to zero:

$$\nabla \times \nabla f = 0$$

(c) In the identities that follow, u and v represent scalar functions while A and B represent vector functions. The extent of the operation of the Del Operator is shown by the over-bar:

$$A \times (B \times C) = (C \times B) \times A = B(A \cdot C) - C(A \cdot B)$$

$$\nabla(uv) = u\nabla v + v\nabla u$$

$$\nabla(A \cdot B) = A \times (\nabla \times B) + (A \cdot \nabla)B + B \times (\nabla \times A) + (B \cdot \nabla)A$$

$$\nabla \cdot uA = u\nabla \cdot A + A \cdot \nabla u$$

$$\nabla \cdot (A \times B) = B \cdot \nabla \times A - A \cdot \nabla \times B$$

$$\nabla \times (uA) = u\nabla \times A - A \times \nabla u$$

$$\nabla \times (A \times B) = (B \cdot \nabla)A + A(\nabla \cdot B) - (A \cdot \nabla)B - B(\nabla \cdot A)$$

$$(\overline{\nabla \cdot A)B} = (A \cdot \nabla)B + B(\nabla \cdot A)$$

$$\nabla \times (\nabla \times A) = \nabla(\nabla \cdot A) - (\nabla \cdot \nabla)A$$

It should be noted here that it is common practice to combine multiple operators: curl of the gradient, divergence of the curl, divergence of the gradient, and curl of the curl. These also lead to properties: distributive properties, vector dot product, vector cross product, product of a scalar and vector, and product rule for the gradient. Again, the works of Schey (1997) and Griffiths (1999) are useful on these aspects.

In Vector Calculus, three fields are widely discussed: (1) *Conservative Vector Field*, (2) *Solenoidal Vector Field*, and (3) *Laplacian Vector Field*. What follow are brief descriptions and mathematical notations of these fields.

The *Conservative Vector Field* refers to the gradient of a scalar potential. There are two closely related concepts in this field. The first is the *path independence* vector field, which is the property of every conservative vector. The second is the *irrotational* vector field, which describes the fact that every conservative vector field has a zero curl. A vector field v is characterized as *conservative* if a scalar field φ such that

$$\mathbf{v} = \nabla\varphi.$$

In the preceding equation,

$$\nabla\varphi$$

denotes the gradient of φ. When the equation is validated, φ is called a scalar potential for v. According to the fundamental theorem of Vector Calculus, any vector field can be expressed as the sum of a conservative vector field and a *Solenoidal* field.

The *Slenoidal Vector Field*, also called the *Incompressible Vector Field,* is a vector field v with divergence zero:

$$\nabla \cdot \mathbf{v} = 0.$$

Whenever a vector field v has only a vector potential component, the condition of zero is satisfied. This is because the definition of the vector potential A is as follows:

$$\mathbf{v} = \nabla \times \mathbf{A}$$

This automatically results in the identity that can be shown, for example, by using Cartesian coordinates:

$$\nabla \cdot \mathbf{v} = \nabla \cdot (\nabla \times \mathbf{A}) = 0.$$

Conversely, for any *Solenoidal* v, there exists a vector potential A such that

$$\mathbf{v} = \nabla \times \mathbf{A}.$$

The divergence theorem provides the equivalent integral definition of a *Solenoidal* field: i.e. for any closed surface S, the net total flux through the surface must be zero:

$$\iint_S \mathbf{v} \cdot d\mathbf{s} = 0$$

where *ds* is the outward normal to each surface element.

The *Laplacian Vector Field* is both *irrotational* and *incompressible*. When the field is denoted as v, it is described by the following differential equations:

$$\nabla \times \mathbf{v} = 0,$$

$$\nabla \cdot \mathbf{v} = 0.$$

Since the curl of v is zero, v is expressed as the gradient of a scalar potential φ:

$$\mathbf{v} = \nabla \phi \qquad (1)$$

And because the divergence of v is also zero, it follows from equation (1) that

$$\nabla \cdot \nabla \phi = 0$$

which is equivalent to

$$\nabla^2 \phi = 0 .$$

Consequently, the potential of a *Laplacian* field satisfies Laplace's equation.

AFRICAN APPLICATIONS

A painstaking search, since even the word Vector is not in their indices, led us to only two works that have applied Vector Calculus in the study of African designs. These works are those of Paulus Gerdes (1999) and Ron Eglash (1999).

In his book titled *Geometry from Africa: Mathematical and Educational Explorations* (1999), Gerdes examines several possibilities for an educational use of two methods for the construction of the rectangular bases used in house building among the Mozambican peasantry. He begins by suggesting the formulation of alternative rectangle axioms and then links them with Vector Geometry to investigate the possible generalizations and variations both in terms of place and space.

According to Gerdes, most Africans in Sub-Saharan Africa traditionally build houses with circular or rectangular bases. He notes that among the Mozambican peasantry, the following two methods are commonly used for the construction of rectangular houses (1999:94-95):

M1 The house builders start by laying down on the floor two long bamboo sticks of equal length. Then these first two sticks are combined with two other sticks also of equal length, but normally shorter than the first ones. Now the sticks are moved from the closure of a quadrilateral. The figure is further adjusted until the diagonals—measured with a rope—become equal. Then, from where the sticks are now lying on the floor, lines are drawn and the house builders can start.

M2 The house builders start with two ropes of equal length that are tied together at their midpoints. A bamboo stick, whose length is equal to that of the desired width of the house, is laid down on the floor and at its endpoints pins are hit into the ground. An endpoint of each of the ropes is tied to one of the pins. Then the ropes are stretched and at the remaining two endpoints of the ropes, new pins are hit into the ground. These four pins determine the four vertices of the house to be built.

Gerdes points out that the knowledge behind these rectangle constructions is equivalent to the following theorems in Euclidean Geometry (1999:95):

T1 A parallelogram with congruent diagonals is a rectangle.

T2 A quadrilateral with congruent diagonals that intersect at their midpoints is a rectangle.

Looking for possibly interesting didactic alternatives of axiomatic constructions for Euclidean Geometry, Gerdes suggests the rectangle axiom proposed by Soviet mathematician, physicist, philosopher and mountaineer Aleksandr Danilovich Alexandrov (Gerdes, 1999:95):

RA If in a quadrilateral $ABCD$, $AD = BC$ and $\llcorner A$ and $\llcorner B$ are right angles, then $AB = DC$ and $\llcorner C$ and $\llcorner D$ are also right angles.

Thus, according to Gerdes, the knowledge underlying the traditional Mozambican house building techniques might be employed to formulate the following alternative rectangle axioms (1999:96):

RA1 If in a quadrilateral $ABCD$, $AD = BC$, $AB = DC$ and $AC = BD$, then $\llcorner A$, $\llcorner B$, $\llcorner C$ and $\llcorner D$ are right angles.

RA2 If in a quadrilateral $ABCD$, $M = AC \cap BD$ and $AM = BM = CM = DM$, then $\llcorner A$,$\llcorner B$, $\llcorner C$, and $\llcorner D$ are right angles, $AB = BC$ and $AB = DC$.

Gerdes then considers the transformation of the Mozambican traditional house techniques into a rectangular block in space of a parallelepiped by adjusting its four diagonals until they become equally long. The result is a hexahedron. Similarly, by adjusting three diagonals to become equally long, the result is an octahedron (Gerdes, 1999:97).

Finally, Gerdes translates the knowledge involved in the two methods for the construction of the rectangular bases of the Mozambican traditional houses into the language of Vector Geometry and Linear Algebra to give a sufficient and necessary condition for the perpendicularity of two vectors. In the case of the first construction method (M1), the following is yielded (Gerdes, 1999:98):

P1 $|p + q| = |p - q/ \leftrightarrow p \perp q$, where p and q represent vectors.

In the second case (M2), the following is the result (Gerdes, 1999:99):

P2 $|r| = |s| \leftrightarrow (r + s) \perp (r - s)$, where r and s represent vectors.

Eglash in his book, *African Fractals: Modern Computing and Indigenous Design* (1999), explores the funeral ritual of the Jola, an ethnic group that cross-cuts the boundaries of Senegal, The Gambia, and Guinea-Bissau. Eglash describes the ritual as follows:

> The body of the deceased was placed on a platform, and posts at each of the four corners are held aloft by pallbearers. If critical knowledge is thought to have been held by the deceased (e.g., as in the case of a murder), a priest asks questions. The pallbearers, reacting to the force of the deceased, move the platform to the right for yes, left for no, and forward for "unknown" (1999:184).

Eglash simulates this ritual on an analog feedback network. He makes no assumptions about whether the pallbearers are exerting force due to conscious opinions or subconscious beliefs because he believes that it is only necessary to assume that they exert force in proportion to this motivation. This is because, for Eglash, if they can both exert force and sense it from others, this would theoretically allow the summation of knowledge among the participants to be expressed in the most effective way possible. In fact, he argues, the technique is more effective than a vote, since voting can lead to the paradox of a minority opinion win if there are more than two options. Besides, according to Eglash, the information emerged from the bottom-up interaction of the parts, yet it was also intentional in the sense that this mechanism for aggregate self-organization of knowledge had been consciously designed. For him, this was not intentionality as he knew it; instead, it sounded more like a description of a neural network in computer science as described by Michael Waldrop:

> If a programmer has a neural network model of vision, for example, he or she can simulate the pattern of light and dark falling on the retina by activating certain input nodes, and then letting the activation spread through the connections into the rest of the network. The effect is a bit like sending shiploads of goods into a few cities along the seacoast, and then letting a zillion trucks cart the stuff along the highways among the inland cities. But if the connections have been properly arranged, the network will soon settle into a self-consistent pattern of activation that corresponds to a classification of the scene. "That's a cat!" (Waldrop, 1992:289-290 cited in Eglash, 1999:165-166).

From the perspective of Vector Calculus, Eglash describes the simulation the following way (1999:165):

(a) In the Jola ritual four pallbearers hold a platform aloft and move it in response to questions. Since the information (whether one believes it to be of spiritual or mundane origin) is held by the pallbearers, we can model the force of each corner as having direction and magnitude (a vector) determined by the pallbearer's conviction. Decision making based on a continuous range rather than on yes/no is called "fuzzy logic" in mathematics.

(b) We can think of the information processing in the Jola funeral as the equivalent of a neural net in which the sum of the force vectors of all four pallbearers are inputs to three amplifiers, with each inverted output connected as negative feedback to the other two. This would require pallbearers to both exert force as well as sense it, but such force-feedback is actually quite common in motor tasks.

From this simulation, it is quite clear that one key mechanism in complexity theory is memory, as the theory postulates that self-organizing systems will utilize 1/F distributions in memory length.

CONCLUSION

Evident in the preceding discussion is the fact that studying African designs using Vector Calculus calls for taking into account the handedness of the coordinate system. Most of the analytic results are easily understood in a more general form using this subset machinery of differential geometry that is Vector Calculus.

Indeed, the making of designs through aggregate self-organization, while unlike planned structures, do not appear to be the result of unconscious social dynamics that we observe for the urban sprawl of Western cities. This may be the result of the difference between African concepts of intention, which can apply to a group project developed over several generations, as opposed to the Western focus on an individual performing immediate action in defining intentionality. What is vital is that there are indications that this pattern creation by Africans through group activity is supported by conscious mechanisms specific to self-organization as conceptualized in complexity theory. At least some graphic counterparts of both the scaling distribution of interactions with memory and the spectrum from order to disorder that exist in European designs can be found in African designs as well. Consequently, the combination of negative and positive feedback appears to be the best candidate for a conscious mechanism.

Chapter Nine

The Fourier Transform and African Applications

INTRODUCTION

In this chapter, we examine the Fourier Transform and its African applications. We begin by providing some background on the Fourier Transform, including its multiple definitions, the theory that underlies it, a sample of its properties, and its domain and range. After that, we discuss the applications of the Fourier Transform to African designs. This is very important to demonstrate the mathematical complexity and intentionality entailed in African designs. In the end, a conclusion is drawn.

This chapter is important because at least two critical questions can be raised concerning works that employ the Fourier Transform to study African designs: (1) Since some scholars have found that all cities (historic, primitive and modern) are fractal precisely because they are complex natural systems, and other scholars have discovered that fractal tiling patterns exist on some of the oldest European tiled floors and in ancient Chinese art, what then does this say for the validity of arguments concerning African fractals? (2) At what number of scales does self-similarity occur in African fractals and what method can a researcher employ to determine self-similarity? These questions are addressed in the following paragraphs.

First, a researcher can demonstrate that traditional African settlements typically show repetitions of similar patterns at ever-diminishing scales: circles of circles of circular dwellings, rectangular walls enclosing ever-smaller rectangles, and streets in which broad avenues branch down to tiny footpaths with striking geometric repetition. The researcher can easily identify the fractal structure when s/he compares aerial views of African villages and cities with corresponding fractal graphics simulations. To estimate the fractal dimension of a spatial pattern, the researcher can use at least two approaches: (1)

in a black-and-white architectural diagram, a two-dimensional version of the ruler size versus length plots can be employed; (2) for an image in shades of gray, the Fourier Transform can be used. Nonetheless, we cannot just assume that African fractals show an understanding of fractal geometry, nor can we dismiss that possibility. Thus, we must listen to what the designers and users of these structures have to say about them. This is because what may appear to be an unconscious or accidental pattern might actually have an intentional mathematical component.

Second, it has been shown that in African designs and knowledge systems, five essential components (recursion, scaling, self-similarity, infinity, and fractional dimension) keep a researcher on track of what does or does not match fractal geometry. Since scaling and self-similarity are descriptive characteristics, the first step is to look for the properties in African designs. Once one establishes that theme, s/he can then ask whether or not these concepts have been intentionally applied and then starts to look for the other three essential components. One can then look for the clearest illustrations of indigenous self-similar designs in African architecture.

THE FOURIER TRANSFORM: SOME BACKGROUND

The Fourier Transform is well covered in the field of Mathematics and other sciences, leading to the production of many works on the subject. To avoid redundancy in citing works employed for the discussion in this section, we will simply state that what follows was gleaned from the following sources which are fully cited in bibliography: Bracewell (2000), Brigham (1988), Cohen and Ryan (1995), Folland (1992, 1999), Gasquet and Witomski (1999), Halmos (1964), Hubbard (1996), Jackson (1963), James (995), Kammler (2000), Körner (1993), Nussbaumer (1982), Morrison (1994), Oberhettinger (1973), Papoulis (1962), Polyanin and Manzhirov (1998), Ramanathan (1998), Ramirez (1985), Ruskai et al. (1992), Sneddon (1995), Sogge (1993), Spiegel (1974), Stein and Weiss (1971), Tolstov (1976), Walker (1996), Wickerhauser (1994), and Yoshida (1968).

Employed in a wide range of applications, such as image analysis, image filtering, image reconstruction, and image compression, the Fourier Transform is generally defined as a vital image processing tool used to decompose an image into its sine and cosine components. The input image is called the spatial domain equivalent, and the output of the transformation represents the image in the Fourier or frequency domain. In the Fourier domain, each point is a representation of a particular frequency contained in the spatial domain image.

A number of common conventions have also been developed over to time to define the Fourier Transform of a more complex-valued Lebesgue *integrable* function f:R→C. One common definition is the following:

$$F(\nu) = \int_{-\infty}^{\infty} f(t)\, e^{-i2\pi\nu t}\, dt,$$

for every real number ν.

When the independent variable t represents *time* (with SI unit of seconds), the transform variable ν represents ordinary frequency (in Hertz). If f is Hölder continuous, then it can be reconstructed from F by the *inverse transform* as follows:

$$f(t) = \int_{-\infty}^{\infty} F(\nu)\, e^{i2\pi\nu t}\, d\nu,$$

for every real number t.

Other notations for $F(\nu)$ are

$$\hat{f}(\nu),\ \mathcal{F}\{f(t)\},\ \mathcal{F}\{f\}(\nu), (\mathcal{F}f)(\nu).$$

The interpretation of the complex function $F(\nu)$ can be aided by expressing it in polar coordinate form: $F(\nu)=A(\nu)e^{i\varphi(\nu)}$ in terms of the two real functions $A(\nu)$ and $\varphi(\nu)$, where

$A(\nu)=|F(\nu)|$, is the amplitude and
$\varphi(\nu)=\arg(F(\nu))$, is the phase.

Then the inverse transform can be written as follows:

$$f(t) = \int_{-\infty}^{\infty} A(\nu)\, e^{i(2\pi\nu t+\phi(\nu))}\, d\nu,$$

which is a recombination of all the *frequency components* of $f(t)$. Each component is a complex sinusoid of the form e $2\pi i\nu t$ whose amplitude is $A(\nu)$ and whose initial phase angle is (at t =0) is $\varphi(\nu)$.

The Fourier transform is often written in terms of angular frequency: $\omega = 2\pi\nu$ whose units are radians per second. The substitution $\nu = \omega/(2\pi)$ into the preceding formulae produces this convention:

$$F(\omega) = \int_{-\infty}^{\infty} f(t)\, e^{-i\omega t}\, dt$$

$$f(t) = \frac{1}{2\pi} \int_{-\infty}^{\infty} F(\omega)\, e^{i\omega t}\, d\omega,$$

which is also a bilateral Laplace Transform evaluated at s=iω. The 2π factor can be split evenly between the Fourier Transform and the inverse, which leads to the following popular convention:

$$F(\omega) = \frac{1}{\sqrt{2\pi}} \int_{-\infty}^{\infty} f(t)\, e^{-i\omega t}\, dt$$

$$f(t) = \frac{1}{\sqrt{2\pi}} \int_{-\infty}^{\infty} F(\omega)\, e^{i\omega t}\, d\omega.$$

This makes the transform *unitary*.

Variations of the preceding conventions can be created by conjugating the complex-exponential kernel of both the forward and the reverse transform. The only requirement is that the signs must be opposites. Other than this requirement, the choice of any of them hinges on convention.

Theoretically, the Fourier Transform postulates that since image is only defined on a closed and bounded domain (the image window), the image can be defined as being zero outside the window. Stated differently, the image function is *integrable* over the real line.

To see how the Fourier Transform works, a researcher begins with a one-dimensional signal and considers a simple step function. This is equivalent to taking a horizontal slice through an image that is black on its left half and white on its right half. A step function (or a square waveform) can then be represented as a sum of sine waves of frequency: ω, 3ω, 5ω, where ω is the frequency of the square wave, recalling that frequency = 1/wavelength. Normally, frequency is the rate of repetitions per unit: i.e. the number of cycles per second, or Hertz. In images, a researcher is concerned with spatial frequency: i.e. the rate at which brightness in the image varies across the image, or varies with the viewing angle.

From the decomposition of the signal in varying sinusoidal components, a researcher can construct a diagram displaying the amplitudes of all the sinusoids for all the frequencies. The researcher has to take into account negative frequencies, so that the sinusoidal component of frequency *f* and amplitude

A_1 has to be split into two components of amplitude $A_1/2$ at the frequencies $+f$ and $-f$. A graph of the amplitude of the Fourier components is called the *spectrum* of the waveform.

Theoretically, then, the Fourier Transform means that when a researcher calculates the Fourier Transform of an image, s/he treats the intensity signal across the image as a function, not just an array of values. In essence, the Fourier Transform describes a way of decomposing a function into a sum of orthogonal functions in just the same way as one decomposes a point in Euclidean space into the sum of its vector components.

The following are the commonly referenced Fourier Transform properties. The many other properties include autocorrelation, discrete function, fast Fourier, Fourier series, Stieltjes transform, Fourier Transform-1, cosine, delta function, exponential function, Gaussian Fourier transformation, Heaviside step formation, inverse function, Lorentzian function, ramp function, rectangle function, traditional function, Hankel transform, Hartley transform, integral transform, Laplace transform, structural factor, Wiener-Khinchin theorem, Winograd transform, completeness, multi-dimensional version, Plancherel theorem, localization property, eigenfunctions, analysis of differential equations, convolution theorem, cross-correlation theorem, tempered distributions, functional relationships, square-integrable functions, distributions, two-dimensional functions, and three-dimensional functions. Due to space limitations, these other properties cannot be discussed in this paper. The interested reader can learn about them by consulting the sources cited in the references.

Where

$$f(t) \quad \stackrel{\mathcal{F}}{\Longleftrightarrow} \quad F(\omega)$$

denotes that $f(t)$ and $F(\omega)$ are a Fourier transform pair.

Linearity:

$$a \cdot f(t) + b \cdot g(t) \quad \stackrel{\mathcal{F}}{\Longleftrightarrow} \quad a \cdot F(\omega) + b \cdot G(\omega)$$

Multiplication:

$$f(t) \cdot g(t) \quad \stackrel{\mathcal{F}}{\Longleftrightarrow} \quad \frac{1}{\sqrt{2\pi}} \cdot (F * G)(\omega)$$ (unitary normalization convention)

$$\stackrel{\mathcal{F}}{\Longleftrightarrow} \quad \frac{1}{2\pi} \cdot (F * G)(\omega)$$ (non-unitary convention)

$$\overset{\mathcal{F}}{\Longleftrightarrow} \quad (F * G)(\nu) \qquad \text{(ordinary frequency)}$$

Modulation:

$$f(t) \cdot \cos \omega_0 t \quad \overset{\mathcal{F}}{\Longleftrightarrow} \quad \frac{1}{2}[F(\omega + \omega_0) + F(\omega - \omega_0)], \qquad \omega_0 \in \mathbb{R}$$

$$f(t) \cdot \sin \omega_0 t \quad \overset{\mathcal{F}}{\Longleftrightarrow} \quad \frac{i}{2}[F(\omega + \omega_0) - F(\omega - \omega_0)]$$

$$f(t) \cdot e^{i\omega_0 t} \quad \overset{\mathcal{F}}{\Longleftrightarrow} \quad F(\omega - \omega_0)$$

Convolution:

$$(f * g)(t) \quad \overset{\mathcal{F}}{\Longleftrightarrow} \quad \sqrt{2\pi} \cdot F(\omega) \cdot G(\omega) \quad \text{(unitary convention)}$$

$$\overset{\mathcal{F}}{\Longleftrightarrow} \quad F(\omega) \cdot G(\omega) \qquad \text{(non-unitary convention)}$$

$$\overset{\mathcal{F}}{\Longleftrightarrow} \quad F(\nu) \cdot G(\nu) \qquad \text{(ordinary frequency)}$$

Integration (example of convolution):

$$\int_{-\infty}^{t} f(\tau)\, d\tau = (f * u)(t) \quad \overset{\mathcal{F}}{\Longleftrightarrow} \quad \frac{1}{i\omega} F(\omega) + \pi F(0) \cdot \delta(\omega),$$

Conjugation:

$$\overline{f(t)} \quad \overset{\mathcal{F}}{\Longleftrightarrow} \quad \overline{F(-\omega)}$$

Scaling:

$$f(at) \quad \overset{\mathcal{F}}{\Longleftrightarrow} \quad \frac{1}{|a|} F\left(\frac{\omega}{a}\right), \qquad a \in \mathbb{R}, a \neq 0$$

Time reversal:

$$f(-t) \quad \overset{\mathcal{F}}{\Longleftrightarrow} \quad F(-\omega)$$

Time shift:

$$f(t - t_0) \quad \overset{\mathcal{F}}{\Longleftrightarrow} \quad e^{-i\omega t_0} \cdot F(\omega)$$

Parseval's theorem:

$$\int_{-\infty}^{\infty} f(t) \cdot \overline{g(t)}\, dt = \int_{-\infty}^{\infty} F(\omega) \cdot \overline{G(\omega)}\, d\omega \qquad \text{(unitary convention)}$$

$$= \frac{1}{2\pi} \cdot \int_{-\infty}^{\infty} F(\omega) \cdot \overline{G(\omega)}\, d\omega \quad \text{(non-unitary convention)}$$

$$= \int_{-\infty}^{\infty} F(\nu) \cdot \overline{G(\nu)}\, d\nu \qquad \text{(ordinary frequency)}$$

In terms of its domain and range, the Fourier Transform cannot be described as two well-defined sets of functions. Instead, the functions can be chosen in several different ways based on what is meant by a *function* and an *integral*. Also, for some pair of domain and range that can be described for the Fourier Transform, it may sometimes be of interest to consider the restriction of the transform to a proper subset of the domain. In general, however, it is imperative to describe as large sets as possible for the domain. The discussion of such extensions, which can be done in different ways and might lead to domains in which either one is not a subset of the other, is beyond the scope of this paper.

AFRICAN APPLICATIONS

To the best of our knowledge, Ron Eglash and his colleagues (1995, 1997 & 1999) are the only researchers that have applied the Fourier Transform to the study of African designs. Thus, the discussion in this section is based on these scholars' work.

In his essay titled "African Influences in Cybernetics" (1995), Eglash discusses the problems of natural/artificial dualisms encountered by cyborgs by arguing that they are similar to those which plague activists and theorists in the long historical battles against racism. He states that "Primitivist" racism operates by making non-Western culture too concrete and, therefore, "closer to nature": i.e. not really a culture at all, but rather beings of uncontrolled emotion and direct bodily sensation, rooted in edenic ecology (an ecology in which both animals and plants are plentiful). He adds that "Orientalist" racism operates by making non-Western culture too abstract and, thus, "arabesque": i.e. not really a "natural" human, but one devoid of emotion, caring only for money and an inscrutable spiritual transcendence. According to him, racism on the African continent, tending towards Orientalism in the North and Primitivism in the South, precludes any simple opposition that a category like "African cybernetics" might hold. Thus, he argues that an anti-racist characterization of African influences in cybernetics must be situated in ways that do not merely reverse or refute its claims, but addresses its historical construction. He observes that opposition to racism has often been composed through two totalizing, essentialist strategies: (1) sameness and (2) difference. He cites Valentin Mudimbe who demonstrates how the category of a singular "African philosophy" has been primarily an invention of difference, having its creation in the play between "the beautiful myths of the 'savage mind' and the African ideological strategies of otherness." Eglash also cites the contrasting perspective of structuralists such as Claude Lévi Strauss who have attempted to prove that African conceptual systems are fundamentally the same as those of Europeans (both having their basis in arbitrary symbol systems). According to Eglash, the problem of these unitary assessments of epistemological status is made particularly clear by the contradictions in the philosophical approach of Sandra Harding, in which African conceptual views were at first characterized as the holistic opposite of Western reductionism, and then soon after as having exactly he same analytic approach as Western science. But as Eglash reminds us, Mudimbe notes that neither sameness nor difference will suffice (Eglash, 1995).

For Eglash, this critique indicates that the analysis of interactions between cybernetic theory and the African Diaspora should not be limited to a purely epistemological perspective. At the same time, however, he argues, socially grounded analyses of science have all too often presented a kind of "Realpolitik" approach to the social construction of cybernetics, one in which the science of computation and control systems is merely a thin disguise for methods of social domination and control. Here, according to Eglash, any subaltern identity (female, non-white, working class, etc.) appears only as yet another powerless victim, and typically one for whom a previously natural existence is endangered by the intrusion of artifice. Thus, for him, the focus on African contributions to cybernetics should not be an attempt to overlook the brutal tragedies enacted by that science, but rather to underscore the multifaceted aspects of its history and the attendant possibilities for resistance and reconfigurations. By moving between questions of epistemological structure and social constructions of science, he believes, suggests some possible origins of cybernetic theory in African culture, ways that Black people have negotiated the rise of cybernetic technology in the West, and the confluence at these histories in the lived experience of the African Diaspora (Eglash, 1995).

Eglash tells us that cybernetic theory is based on two dimensions of communication systems: (1) the information structure and (2) the physical representation of that information. He says that the most fundamental characteristic of an information structure is its computational complexity, which is a measure of its capacity for recursion (i.e. self-reference, reflexivity). This mathematical result agrees nicely with our intuition about the crucial role of reflexive awareness in our own "information structure." The most fundamental characteristic of a representational system is the analog-digital distinction. Digital representation requires a code table (the dictionary, Morse code, the genetic code, etc.) based on physically arbitrary symbols (text, numbers, flag colors, etc.). Eglash equates this postulate with Ferdinand de Saussure's "arbitrariness of the linguistic signifier." Eglash adds that analog representation is based on the proportionality among the physical changes in a signal and changes in the information it represents (e.g., waveforms, images, vocal intonation). An example he cites is that as a person's excitement increases, so does the loudness of his/her voice. Similarly, while digital systems use grammars, syntax, and other relations of symbolic logic, analog systems are based on physical dynamics: i.e. the realm of feedback, hysteresis, and resonance. This dichotomy, according to him, is fundamental to current cybernetic debates concerning, for example, which type of representation is used by neurons in the human brain, or the type recommended for artificial brains (Eglash, 1995).

Eglash recounts that in the early years of American cybernetics, analog and digital systems were seen as epistemologically equivalent, both considered capable of complex kinds of representation. But by the early 1960s, a political dualism was coupled to this representation dichotomy. Eglash points out that the "counterculture" radicals of the cybernetics community—Norbert Wiener, Gregory Bateson, Hazel Henderson, Paul Goodman, Kenneth Boulding, Barry Commoner, Margaret Mead, among others—made the erroneous claim that analog systems were more concrete, more "real" or "natural" and, therefore, according to this romantic cybernetics, ethically superior. In social domains, according to Eglash, this perspective converged with Jean Jacques Rousseau's legacy of the moral superiority of oral over literate cultures (Eglash, 1995).

According Eglash, for African Americans, this development meant a debilitating valorization. They could use this ethical claim to combat some racism, albeit only in terms of identifying as unconscious, innocent natives in a lost past. Thus, African modes of representation in the use of sculpture, movement and rhythm were often abandoned to modernist claims that Africa was the culture of non-representation, the culture of the Real. By the 1970s, he notes, widespread epistemological critiques of realism, noting that it is representation that allows self-consciousness and intentionality, resulted in interpretations which limited cultural analysis to arbitrary signifiers. African dance, for example, would be a set of movement symbols, not a waveform (Eglash, 1995).

Subsequently, recounts Eglash, African cultural analysis became split between those who retained the modernist trope of African identity grounded in naturalist realism (recognizing analog systems but refusing to see them as representation), versus those who adopted a postmodern trope of textual metaphor (which avoids primitivism at the expense of abandoning recognition of analog systems): reggae versus rap (Eglash, 1995).

Eglash observes that postmodern cybernetics, however, has shown that analog systems are capable of the flexible representation required to perform complex (Turing Machine-equivalent) computations. He further mentions that in particular, a new appreciation for analog systems was fundamental to the rise of fractal geometry, nonlinear dynamics, and other branches of chaos theory. He notes that by viewing physical systems as forms of computation, rather than merely inert structures, researchers became open to the possibility of having infinite variation in deterministic physical dynamics. Thus, for him, analog systems can achieve the same levels of recursive computations as digital systems, as the two are epistemologically equal (Eglash, 1995).

Put differently, Eglash believes that the appeal to digital systems in African culture may well have been a necessary antidote to the skewed social portrait of it, but it is not the only recourse for combating ethnocentric epistemological claims. He asserts that African cultures pave indeed developed systems of analog representation which are capable of the complexities of recursion, and there are indications that this indigenous technology has been in conversation with cybernetic concepts in the West (Eglash, 1995).

In terms of Africa in the origins of cybernetics, Eglash provides abundant evidence. First, the use of African material culture as a form of analog representation is particularly vivid in cases of recursive information flow. In African architecture, he cites the fact that recursive scaling—i.e. fractal geometry—can be seen in a variety of forms. In North Africa, he says that it is associated with the feedback of the "arabesque" artistic form, particularly in the branches of branches forming city streets. In Central Africa, he states that it can be seen in additive rectangular wall formations, and in West Africa one can see circular swirls of circular houses and granaries. He argues that this is not limited to a visual argument, as the fractal structure of African settlement patterns has been confirmed by computational analysis of digitized photos in earlier work with his colleagues (Eglash, 1995).

Second, Eglash cites the recursive scaling in Egyptian temples that can be viewed as a formalized version of the fractal architecture found elsewhere in Africa, and it is most significant in the use of the Fibonacci sequence. The sequence is named for Leonardo Fibonacci (ca. 1175–1250), who is also associated with an unusual example of recursive architecture in Europe. The Fibonacci sequence was one of the first mathematical models for biological growth patterns, and it inspired Alan Turing and other important figures in the history of computational morphogenesis. Since Fibonacci was sent to North Africa as a boy and devoted his years there to mathematics education, Eglash argues that it is possible that Fibonacci's seminal example of recursive scaling is of African origin (Eglash, 1995).

Third, Eglash mentions that Benoit Mandelbrot, dubbed the "father of fractal geometry," reports that his invention is the result of combining the abstract mathematics of Georg Cantor with the empirical studies of H. E. Hurst. Cantor was a 19th-Century Rosicrucian mystic, who often combined his mathematics with his religious belief. His cousin Moritz Cantor was a famous scholar in the geometry of Egyptian art and architecture. According to Eglash, given these facts, and the similarity of this first European fractal to the Egyptian architectural structure symbolizing creation (the lotus), an Egyptian origin is likely here as well (Eglash, 1995).

Fourth, Eglash cites Arnold Goldsmith who reported golem legends going back to the 4th Century BCE and described their continuing popularity in Jewish legend. Norbert Weiner, the Jewish dean of analog cybernetics, was quite influenced by this concept of information embedded in physical dynamics. Eglash notes that Weiner made several references to the golem in his writing and reported that even as a child he was fascinated by the idea of making a doll come alive. His religious identity was closely tied to *gashmuit*, the informal, physical (and traditionally female) side of Judaism, and he was particularly proud of his ancestry to famed Egyptian physician Moses Maimonides (Eglash, 1995).

Fifth, Eglash points out that in addition to spatial analog representation, many African societies have developed techniques for the analog representation of time-varying systems, including transformation into frequency or phase-domain representation. He mentions the fact that animist energy flow, drawn for him by a Bambara seer, can be visualized as a sort wave emanating from a sacrificial egg. The dashed lines inside the figure are a digital code symbolizing good fortune. He notes that undulatory schemes in Egyptian art that show an understanding of motion as a rhythmic time series and the transformation of time series to a frequency-domain representation can be seen in African conceptualizations of circular time. According to him, the extreme in African time series analysis is the search for patterns in the Nile floods. He adds that the data set taken once a year for 15 centuries became the basis for the work of H. E. Hurst. A British civil servant, Hurst spent 62 years in Egypt, and finally deduced a scaling law, based on this time series, which Mandelbrot used to bring Cantor's abstract set theory into empirical practice (Eglash, 1995).

Finally, Eglash states that the most common frequency analysis used by Norbert Wiener and others in modern cybernetics is the Fourier Transform. Egalsh notes that Joseph Fourier began his work with an analysis of René Descartes' equations; Fourier did not leave this static framework until his expedition to Egypt in 1798, where he analyzed the geometry of Egyptian architecture. It was in Egypt that he devised the basis for the Fourier Transform. A comparison of Fourier's visualizations of convergence of a sequence with a diagram of Egyptian architecture (which, because of the Fibonacci sequence, also shows convergence to a limit), suggests that the African concept of recursive structure and dynamic form may have contributed to this analysis as well (Eglash, 1995).

As for Black cybernetics in the postmodern era, Eglash raises the following question: Setting aside both the definition of cybernetic and its interaction with popular culture, what kinds of technological capability does the vernacular cybernetics of the African American community represent? He responds by stating that one clear illustration can be found in the striking utilization of the analog/digital dualism for the production of musical signifiers in the divisions between reggae and rap music. He notes that reggae is more aligned with the naturalizing trope modernity, and rap with the artificial affinities of the postmodern. He points out that in reggae, the language of analog representation abounds: as "Rastaman Vibration" lets us "tune into de riddem," we become resonant nodes linked by the waveforms of a polyphonic beat. In rap music, he observes, it is the digital communication that signifies cultural identity; natural harmonies are broken up by arbitrary sound bites and vocal collage, and the melody is subordinated to a newly spliced code: a mutant reprogramming of the social software (Eglash, 1995).

According to Eglash, from the viewpoint of cultural studies, the utilization of the analog/digital division in reggae versus rap does indeed count as a technological capability. But, he asks, would it also count from the view of a cybernetics engineer? His response is that the use of the scratch sound is associated with the birth of Rap, but phonograph records are analog devices. Similarly, reggae makes use of an army of both analog and digital audio equipment. In responding to whether the use of technological language by African Diasporic subcultures is merely linguistic play, Eglash says no. He believes that despite (in fact because of) the wide assortment of apparatus, rap and reggae artists have created a technology for signal processing that would indeed meet the specificities of current cybernetics engineering. He states that the evidence for this begins in the work of Richard Voss, who first measured the fractal dimension for various types of acoustic communication in 1977. Voss discovered that, on the one hand, the physical arbitrariness of digital signifiers meant that the waveforms of digital commutation were a succession of fairly random signals, overall creating a "white-noise spectrum." In analog waveforms, on the other hand, long-term changes in information were reflected in long-term signal changes. Eglash reasons that since there were similar information changes on many scales, the result was a fractal structure, a "1/F noise spectrum," in the case of analog communication. Thus, according to Eglash, the waveform created by pitch changes in speech, which are primarily due to the phonetic differences between words, tends toward a white-noise spectrum, while the pitch signal of music shows the fractal structure of analog representation (Eglash, 1995).

Eglash also notes that Voss later showed that this relationship held for all types of music, both instrumental and vocal, with samples ranging from Indian ragas to Russian folksongs. Egalsh further cites his own studies that show that while reggae music also has this fractal structure, rap is the only music (aside from avant-garde experiments such as those of John Cage) which violates this rule. According to Eglash, the reason for this is the intentional violation of analog representation by digital coding, a violation that invokes rap artists' oppositional stance, but also offers a positive outlook in the possibilities for their cybernetic innovation. Moreover, the rap-reggae fusions that are now becoming increasingly popular (e.g. ragamuffin) have characteristics which indicate that their signals are likely to avenge a fractal dimension value half-way between the two. This precision of control over an abstract cybernetic principle, Eglash argues, indicates that it is not simply a matter of the adaptation of terminology, but that African Diasporic identity is expressed in these examples through a conscious manipulation of complex signal characteristics (Eglash, 1995).

In their article, "Indigenous Science for Education and Development: A Boot-strapping Approach" (1997), Ron Eglash, Egondu Onyejekwe, Christian Sina Diatta and Nfally Badiane discuss their visual observations that led them to find that aerial photos of traditional African settlements tend to have a fractal structure (scaling in street branching, recursive rectangular enclosures, circles of circular dwellings, etc.). This was quantitatively confirmed when they applied a 2-dimensional Fourier Transform to digitized photo images to estimate the fractal dimension from the slope of the spectral density function. Subsequent study showed that these architectural fractals result from intentional designs, not simply unconscious social dynamics, and that recursive scaling structures can be found in other areas of African material culture: art, religion, indigenous engineering, and games. These scholars also find that in the design rationales and cultural semantics of many of these geometric figures, as well as in indigenous quantitative systems (additive progressions, doubling sequences, binary recursion) and symbolic systems (iconic symbols for feedback loops, equiangular spirals, infinity), there are abstract ideas and formal structures that closely parallel some of the fundamental aspects of fractal geometry (Eglash et al., 1997).

According to Egalsh and his colleagues, these results agree with recent developments in complex systems theory, which suggest that pre-modern, non-state societies were neither utterly anarchic, nor frozen in static order, but rather utilized an adaptive flexibility that took advantage of the nonlinear aspects of ecological dynamics. They raise the possibility about how this potent formulation of indigenous knowledge could be applied to problems in education and development. They suggest that one method is to encourage its dissemination in development agencies. They also observe that decades of research have shown that a top-down approach to development, even that making use of indigenous knowledge, is often less effective than a bottom-up, "grass roots" approach. It is their hope that a project designed to create a framework in which to test the possibility that indigenous knowledge can be used in a boot-strapping approach to development. They note that the term boot-strapping is typically applied to computer systems ("boot up the disk"), where a small program is able to self-install a larger system (from the phrase "pulling yourself up by your own bootstraps"). Similarly, they view a boot-strapping approach to development as one which begins with indigenous knowledge under local control, and self-installs modern technological abilities (Eglash et al., 1997).

In his book, *African Fractals: Modern Computing and Indigenous Design* (1999), Eglash discusses his analysis of an aerial photo of Labbazanga, a Songhai village in Mali. The photo's image is in shades of gray. He applies the Fourier Transform to find the scaling slope of 1/F noise in a one-dimensional time series. The method allows him to make a more direct measure of the scaling properties of Labbazanga. He then applies the method to a two-dimensional spatial distribution by sweeping the same spectral density measure around in polar coordinates. Rather than the line of one-dimensional 1/F noise, he employs a two-dimensional distribution characterized by a cone—i.e. 2-D Fourier Transform, with frequency in polar coordinates: wider circle = higher frequency. Since it is difficult to show the entire cone, he takes historical slices which show similar characteristics for both Labbazanga and its fractal simulation (Eglash, 1999:231–233).

Labbazanga shows circular swirls of circular houses without any single focus. A lack of central focus, however, does not imply a lack of self-similarity. Even the decentralized swirls of circular buildings show a scaling symmetry. A quantitative test of the photo confirmed what Eglash's eyes were telling him: the Songhai architecture can be characterized by a fractal dimension similar to that of the computer-generated fractal (Eglash, 1999:31–32).

CONCLUSION

The preceding discussion has demonstrated that African scaling designs should be expected to vary greatly in purpose, pattern, and method. This is because while it is not difficult to invent explanations based on unconscious social forces (for example, the flexibility in conforming designs to material surfaces as expressions of social flexibility), any such explanation can account for a design's diversity.

In addition, it has been shown that from optimization engineering, to modeling organic life, to mapping between different spatial structures, African artisans have developed a wide range of tools, techniques, and design practices based on the conscious application of scaling geometry. Thus, instead of using the Koch curve, for example, to generate the branching fractals used to model the lungs and acacia tree, a researcher can use passive lines that are just carried through the iterations without change, in addition to active lines that create a growing tip by the usual recursive replacement.

Chapter Ten

Mathematical Tiling/Tessellation and African Applications

INTRODUCTION

This chapter is about Mathematical Tiling/Tessellation and how it has been employed to investigate African designs. We start with some background discussion on Mathematical Tiling/Tessellation, including a definition, wallpaper groups, color, quadrilaterals, regularity and irregularity, nature, polygon, vertex, other spaces, surface normal, triangle, graphics, Phong shading, flat shading, and Gouraud shading. Following this is a discussion of some applications of Mathematical Tiling/Tessellation to investigate African designs.

The chapter is vital because it exposes readers to the mathematics of periodic patterns of one or more shapes which can be extended across an entire plane infinitely that are present in certain African designs. As Reza Sarhangi reminds us, some of the most extensive works with mosaic designs were done by the Moorish artists from North Africa. During the 8th Century, a Moorish dynasty invaded Spain and established a civilization in Andalusia that lasted until 1492. The Moorish kings of the 12th and 13th Centuries built the *Alhambra* (a citadel overlooking Granada, Spain) that is the most famous building in the Western world for its tessellation designs. The Moorish tessellation manifested itself in the use of few shapes and colors of tiles to build complex geometric designs. Unlike the extensive use of living beings in the Roman tilings, a practice forbidden (*haram*) by Moorish religious leaders, the works of the Moorish artists were abstract and excluded real-world objects (http://pages.towson.edu). Furthermore, as Tim Helck observes, all 17 distinct types of *plane symmetry groups* (also known as *wallpaper groups*) that have been identified by mathematicians have been found to exist in the decoration of the *Alhambra* (http://www.thelck.com).

TILING/TESSELLATION: SOME BACKGROUND

Tiling, also known as Tessellation (henceforth, the concepts will be used interchangeably as it is common practice), is a well established sub-discipline in Mathematics that has generated many scholarly articles and a number of books. In order not to be redundant in citing works used for the discussion in this section, we will simply state that what follows was delineated from the following works that are cited in the bibliography: Ciucu (2005), Golomb (1994), Gilmer (1998), Grünbaum and Shephard (1987), Krajčheski (2001), Stein and Szabó (1994), and Wieting (1982).

Tiling or Tessellation can be generally defined as the filling up of a two-dimensional space by congruent copies of a figure that do not overlap. The figure that results is referred to as the basic/fundamental shape for the tessellation. One must recall that a regular polygon is a convex polygon whose sides all have the same length and whose angles all have the same measure. Also, a regular hexagon is a regular polygon with six sides. Only two other regular polygons tessellate: (1) the square and (2) the equilateral triangle. Tessellations can be derived by combining translation, reflection (or mirror) and rotation images of the basic shape. Variations of these regular polygons can also tessellate by modifying one side of a regular basic shape and then modifying the opposite side in the same way.

Mathematicians have categorized tilings with translational symmetry into 17 wallpaper groups. As mentioned earlier, all 17 of these patterns exist in the *Alhambra* palace in Granada, Spain.

When discussing a tiling that is displayed in colors, one must specify whether the colors are part of the tiling or just part of its illustration to avoid ambiguity. According to the four-color theorem, every tessellation of a normal Euclidean plane, with a set of four available colors, each tile can be colored in one color such that no tiles of equal color meet at a curve of positive length. It should be noted that the coloring guaranteed by the four-color theorem will not in general be in congruence with the symmetries of the tessellation. To produce a coloring which does, as many as seven colors may be needed. If a parallelogram pattern is colored before tiling it over a plane, seven colors are required to ensure that each complete parallelogram has a consistent color which is distinct from that of adjacent areas. If one tiles before coloring, only four colors are necessary.

A tessellation with 2-fold rotational centers at the midpoints of all sides can be formed by copies of an arbitrary quadrilateral; translational symmetry with as minimal set of translation vectors a pair according to the diagonals of the quadrilateral or, equivalently, one of these and the sum or difference of the two. In terms of an asymmetric quadrilateral, this tiling can be categorized as wallpaper group p2. As a basic domain, it is a quadrilateral. Equivalently, starting from a rotational center, one can construct a parallelogram subtended by a minimal set of translation vectors. As a basic domain, one can divide this by one diagonal and take one half (a triangle: i.e. a three-sided polygon). This type of a triangle has the same area as the quadrilateral and can be constructed from it by cutting and pasting.

A *regular* tessellation is a highly symmetric tessellation comprising congruent regular polygons. Only three regular tessellations have been identified: (1) equilateral triangles, (2) squares, and (3) hexagons. A *semiregular* tessellation comprises a variety of regular polygons; eight of these have been identified by mathematicians. At every vertex point, the arrangement of polygons is identical. An *edge-to-edge tessellation* is even less regular, with the only requirement that adjacent tiles only share full sides: i.e. no tile shares a partial side with another tile. Other types of tessellations exist, depending on types of figures and types of pattern: *regular* versus *irregular*, *periodic* versus *aperiodic*, *symmetric* versus *asymmetric*, and *fractal tessellations*, in addition to other classifications.

The most famous example of tessellations that are aperiodic is the Penrose tiling which uses two different polygons. It belongs to a general class of aperiodic tilings that can be constructed out of self-replicating sets of polygons by using recursion.

A *monohedral* tiling is one in which all tiles are congruent. The Voderberg tiling developed by Hans Voderberg in 1936 is said to be the earliest known spiral tiling, and the unit tile is a bent enneagon. The Hirschhorn tiling was generated by Michael Hirschhorn in the 1970s, and the unit tile is an irregular pentagon. An *edge-to-edge* tiling is topologically identical to a hexagonal tiling, with each hexagon flattened into a rectangle with the long edges divided into two edges. A *basketweave* tiling is topologically identical to the Cairo pentagonal tiling, with one side of each rectangle counted as two edges, divided by a vertex on the two neighboring rectangles.

Tessellations in nature, such as the *Basaltic lava flows,* often display *columnar jointing* due to *contraction* forces that cause cracks as the lava cools. The extensive crack networks that develop often generate hexagonal columns of lava. An example of this is the Giant's Causeway in Ireland. The Tesselated pavement in Tasmania is a rare sedimentary rock formation where the rock has fractured into rectangular blocks.

In contrasting the number of sides of a polygon versus the number of sides at a vertex, for an infinite tiling,

> let a be the average number of sides of a polygon, and b the average number of sides meeting at a vertex, then

$(a - 2)(b - 2) = 4.$

A continuation of a side in a straight line beyond a vertex is counted as a separate side. For a self-repeating tiling, one can take the averages over the repeating part. Generally, the averages are taken as the limits for a region expanding to the whole plane. In the case of an infinite row of tiles, or tiles getting smaller and smaller outwardly, the outside is not negligible and should also be counted as a tile while taking the limit. In extreme cases, the limits may be nonexistent, or they may depend on how the region is expanded to infinity.

For finite tessellations and polyhedra, the following formula is derived:

$$(a - 2)(b - 2) = 4\left(1 - \frac{\chi}{F}\right)\left(1 - \frac{\chi}{V}\right)$$

where F is the number of faces, V is the number of vertices (end points), and χ is the Euler characteristic (for the plane and for a polyhedron without holes). In the plane, the outside is counted as a face. The formula follows observing that the number of sides of a face, summed over all faces, yields twice the number of sides, which can be expressed in terms of the number of faces and the number of vertices. Likewise, the number of sides at a vertex, summed over all faces, yields also twice the number of sides.

In most cases, the number of sides of a face equals the number of vertices of the face, and the number of sides meeting at a vertex equals the number of faces meeting at a vertex. Nonetheless, in a case like two square faces touching at a corner, the number of sides of the outer face is 8; thus, if the number of vertices is counted, then the common corner has to be counted twice. In the same vein, the number of sides meeting at that corner is 4; consequently, if the number of faces at that corner is counted, the face meeting the corner twice must be counted twice.

A tile with a hole, filled with one or more other tiles, is not permissible, since the network of all sides inside and outside is disconnected. It is only permissible with a cut in order for the tile with the hole to touch itself. In counting the number of sides of this tile, the cut must be counted twice.

For Platonic solids, round numbers result because the average over equal numbers is taken:

For $(a - 2)(b - 2)$, one gets 1, 2, and 3.

From the formula for a finite polyhedron, in the case that while expanding to an infinite polyhedron the number of holes (each contributing −2 to the Euler characteristic) grows proportionally with the number of faces and the number of vertices, the limit of (a − 2)(b − 2) is larger than 4. It should be noted that the result does not depend on the edges being line segments and the faces being parts of planes. Despite the mathematical rigor required to deal with pathological cases, they can also be curves and curved surfaces.

In addition to tessellating the 2-dimensional (2D) Euclidean plane, other n-dimensional spaces can be tessellated by filling them with n-dimensional polytopes. Tessellations of other spaces are called honeycombs. Tessellations of other spaces include the following:

(a) Tessellations of n-dimensional Euclidean space. An example is filling 3-dimensional (3D) Euclidean space with cubes to generate a cubic honeycomb.
(b) Tessellations of n-dimensional elliptic space. An example is projecting the edges of a dodecahedron onto its circumsphere to generate a tessellation of the 2D sphere with regular spherical pentagons.
(c) Tessellations of n-dimensional hyperbolic space. An example is Escher's *Circle Limit III* that depicts a tessellation of the hyperbolic plane using the Poincaré disk model with congruent fish-like shapes. The hyperbolic plane permits a tessellation with regular p-gons meeting in q's when

$$\frac{1}{p} + \frac{1}{q} < \frac{1}{2}.$$

Circle Limit III may be depicted as a tiling of octagons meeting in threes, with all sides replaced with jagged lines and each octagon then cut into four fish.

In computer graphics, tessellation techniques are often used to manage datasets of polygons and divide them into suitable structures for rendering. Normally, at least for real-time rendering, the data are tessellated into triangles, a technique that is sometimes referred to as *triangulation*. In computer-aided designs, arbitrary 3D shapes are often too complicated to analyze directly; therefore, they are divided (tessellated) into a *mesh* of small, easy-to-analyze pieces—either irregular *tetrahedrons* or irregular *hexahedrons*. The mesh is utilized for finite element analysis. Some geodesic domes are designed by tessellating the sphere with triangles that are as close to equilateral triangles as possible.

In surface and solid modeling, tiling is used to represent 3D objects as a collection of triangles or other polygons. Both curved and straight surfaces are turned into triangles either at the time they are first generated or in real time when they are rendered. The more triangles utilized to represent a surface, the more realistic the rendering, albeit requiring more computation. Given an object's distance from a camera, triangles may be discarded at the time of their rendering. Some applications yield multiple models with various amounts of triangles and employ the best one depending on distance. The vertices of the triangles are designated X-Y-Z and RGB values that are utilized to compute light reflections for shading and rendering. Tessellation is not employed for 2D graphics, although 2D graphics can be used to draw 3D objects. It is imperative that an artist uses standard drawing tools, color fills and gradients to generate simulation of depth and shading.

In 3D graphics, a surface normal—i.e. an imaginary line that is perpendicular to the surface of a polygon—can be used. It can be computed at the vertex of a triangle as the average of all the vertices of adjoining triangles. It also can be computed for each pixel in the triangle as in Phong shading (more on this later). Surface normals are quite useful in deriving the reflectivity of a light source shining onto an object.

The surfaces of 3D objects are broken down into triangles in 3D graphics. For flat surfaces, small numbers of triangles are utilized; to mold curved surfaces, large numbers of triangles are used in a similar manner that a geodesic dome is constructed. The three points of every triangle (vertices) of an object are to be computed on an X-Y-Z scale and must be recomputed each time that object is moved.

Graphics pipeline—i.e. stages required to transform a 3D image into a 2D screen—is employed in 3D graphics rendering. It is responsible for processing information initially provided similarly to how properties at the end points (vertices) or control points of the geometric primitives are used to describe what is to be rendered. Lines and triangles are the typical primitives in 3D graphics. The types of property provided for each vertex include X-Y-Z coordinates, RGB values, translucency, texture, reflexivity, and other characteristics.

In 3D graphics, flat, Phong and Gouraud shading are the three most common types of shading techniques used. Flat shading is employed to compute a one-tone shaded surface to simulate simple lighting.

Phong shading—a technique developed by Phong Bui-Tuong in the mid-1970s—is used to compute a shaded surface based on the color and illumination at each pixel. The approach is said to be more realistic than Gouraud shading (more on this later), albeit it requires more computation. Phong shading does not produce shadows or reflections. The surface normals at a triangle's points are used to compute a surface normal for each pixel, generating a more accurate RGB value for each pixel.

Gouraud shading—a technique developed by Henri Gouraud in the early 1970s—is used to compute a shaded surface based on the color and illumination at the corners of every triangle. The technique is the simplest rendering method and is computed faster than Phong shading. Like Phong shading, Gouraud shading also does not produce shadows or reflections. But unlike for

Phong shading, for Douraud shading, the surface normals at a triangle's points are used to generate RGB values that are averaged across the triangle's surface.

AFRICAN APPLICATIONS

While one can find many works on earth tessellations dealing with Africa, such as those of Sears (2006) and Sears et al. (2004), not much work has been done on tessellation in African designs. A laborious search yielded a small number of works that have applied the approach to examine African designs, albeit the discussions are not lengthy.

In his study of the Ten-Pointed Star Tessellation in Islamic art, Tim Helck examines this typical pattern in North Africa, in addition to Iran, Iraq, Afghanistan, Pakistan and Southern Spain. He notices that on first glance, the art might seem "impossible," as it is a space-filling pattern based on a pentagon, and the pentagons do not tessellate very well. On closer inspection, however, he notices that in a few key spots, the pentagons are replaced by a six-sided figure that is composed of two overlapping pentagons. This double-pentagon makes it possible for the whole pattern to resolve itself and fit into a rectangular framework (http://www.thelck.com).

Helck finds in the Ten-Pointed Star Tessellation all four types of two-dimensional patterns that can be delineated by "symmetry operations": (1) translation—movement along a straight line, (2) reflection/mirror—matching any axis with its axis on the original, (3) glide reflection—lining up an axis on the original and then sliding the tracing along the axis until the patterns match, and (4) rotation—rotating an axis 180 degrees until it matches the original. His "symmetry operations" yielded three distinct Axes of Translation Symmetry, two distinct Axes of Reflection Symmetry, two distinct Axes of Glide Symmetry, and two distinct Points of two-fold Rotational Symmetry. In the North African variety of the Ten-Pointed Star Patterns, Helck finds that the Pentagon becomes Five-Pointed Stars, the Kites grow into six-sided shapes resembling Shields, and the Ten-Pointed Stars become spinier (http://www.thelck.com).

Gloria Gilmer in her paper, "Mathematical Patterns in African American Hairstyles" (1998), discusses the objective of her work with Black hair geared toward uncovering the Ethnomathematics of some hair braiders and at the same time answer the complex research question about what the hair braiding enterprise can contribute to mathematics education and, conversely, what mathematics education can contribute to the hair braiding enterprise. To her, it is clear that this single practical activity can, by its very nature, generate more mathematics than the application of a theory to a particular case (Gilmer, 1998:1).

Assisted by Stephanie Desgrottes, a 14-year-old student of Haitian descent at Half Hollow Hills East School in Dix Hills, New York, and Mary Porter, a teacher in the Milwaukee Public Schools, Gilmer and her collaborators observed and interviewed hair stylists at work in their salons along with their customers. Their focus was on mathematical tessellations widely used and understood by hair braiders and weavers but not thought of by them as being related to mathematics (Gilmer, 1998:1–2).

Gilmer and her assistants discovered two types of braids that were very common in the salons they visited. One type comprises Box Braids in which the tessellations of boxes are shaped like rectangles and the pattern resembles a brick wall starting with two boxes at the nape of the neck and increasing by one box at each successive level away from the neck. The hair inside the box is drawn to the point of intersection of the diagonals of the box. Braids are then placed at this point. Braids so placed hide the scalp at the previous level in the tessellation. In this style, the scalp is completely hidden. In addition, Gilmer and her assistants were informed that the braids so placed are unlikely to move much when the head is tossed. The other type comprises Triangular Braids in which the triangles are shaped like equilateral triangles. The hair inside the triangle is drawn to the point of intersection of the bisectors of the angles of the triangle. Again, this style allows hair to move less freely than hair drawn to a vertex when the head is tossed (Gilmer, 1998:2).

Paulus Gerdes in his book, *Geometry from Africa: Mathematical and Educational Explorations* (1999), investigates hexagonal weaving among the Makhuwa craftsmen in the north of Mozambique. These craftsmen weave their light transportation baskets (*litenga*) and their fish traps (*lema*) with a pattern of regular hexagonal holes. The strands are woven one-over-one-under in three directions to generate a stable fabric. The same hexagonal basket weaving technique is used in several regions of Africa. It is used in Madagascar to make fish traps and transport baskets, in Kenya to make cooking plates, in Cameroon among the Pygmies to make carrying baskets, in northeastern Congo among the Meje to make covering pots and among the Mangbetu to weave hats (Gerdes 1999:112–113).

Gerdes employs many mathematical approaches to explore the weaving technique. They include tilings, finite designs, geometric functions, and polyhedra. Since this paper is on tessellation and its application on African designs, I will focus on Gerde's application of tilings on the weaving technique. By constructing a piece of the regular hexagonal weaving pattern, Gerdes discovers several tilings of the plane. First, he places the piece on the top of a sheet of paper and then draws the sides of the hexagonal holes. Continuing only with the sheet of paper, he links certain vertices by segments and discovers the regular hexagonal tiling. Once he finds the regular hexagonal tiling, he discovers other tilings by subdivision. And once he finds the tiling with equilateral triangles, he then constructs other polygons. By considering these figures, he formulates the following conjectures (Gerdes, 1999:114):

(a) the sum of the measures of the internal angles of a n-gon is equal to $3(n-2) \times 60°$;
(b) areas of similar figures are proportional to the squares of their sides, the sum of the first n odd numbers is n^2;
(c) the sum of the first n odd numbers is n^2.

In his book, *African Fractals: Modern Computing and Indigenous Design* (1999), Ron Eglash states that while there is no quantitative measure of fractal dimension in pre-colonial African knowledge systems, the idea of a spectrum progressing from more orderly to less orderly, however, is vividly portrayed in certain African material designs. According to him, the best examples of these can be found in the raffia palm textiles of the Bakuba in the Congo. These textiles tend to show periodic tiling along one axis and aperiodic tiling—often moving from order to disorder—along the other. Eglash notes that similar geometric visualizations of the spectrum from order to disorder have been employed in Computer Science. He also points out that the Bakuba weavings never reach more than halfway across the spectrum—they typically move between 1 and 1.5: i.e. from periodic to fractal, rather than stretching all the way to pure disorder. He concludes by stating that the only one African textile he knows of that suggests the full spectrum from order to disorder is the print pattern from West Africa (Eglash, 1999:172–173).

CONCLUSION

Evident from the preceding discussion is that attractive connections with artistic and science education can be constructed by studying tessellations in African designs. Such connections have the potential to increase students' interest and motivation for mathematical activity.

The use of computers and sophisticated software has brought mathematical tessellation to the reach of many individuals and allows them to be creative. Today, students in all grade levels can apply mathematical transformations in a software program such as TesselMania to generate Escher-like designs that would take several days for an artist to generate just a few years ago using pencil drawings.

Chapter Eleven

Bifurcations and African Applications

INTRODUCTION

In this chapter, the focus is on Bifurcations and how they have been utilized to probe African phenomena. We begin by providing some background information on Bifurcations, including general and theoretical definitions of Bifurcation, types of Bifurcations, codimension of a Bifurcation, the Bifurcation Diagram, the Feigenbaum Constants, and Catastrophe Theory. After this, some of the works that have employed Bifurcations to investigate African phenomena are discussed. These works, all articles, appear to be quite recent, dating from the late 1990s to the present.

This chapter is of great import because as Ralph Abraham observes, the concept of Bifurcations is among the five most essential concepts (the other four being States, Trajectories, Attractors, and Basins) from Chaos Theory—a new branch of Mathematics that has evolved into a mass movement and major social transformation dubbed the Chaos Revolution. As a new branch of Mathematics, Chaos Theory, with its attendant concepts, poses a challenge to the orthodoxy of Mathematics, because for a long time there has not been a new branch of Mathematics. And since there is justification for a connection between the mathematical model of chaos and the chaos of everyday life (Abraham, 1996:1), it behooves students of African Mathematics to have some knowledge of Bifurcations.

BIFURCATIONS: SOME BACKGROUND

The theoretical study of Bifurcations, dating to the 1960s, is a relatively younger mathematical phenomenon compared to other subfields of Mathematics, although certain mathematical aspects of Bifurcations can be traced back to ancient Egyptian artefacts (more on this later). Thus, while this sub-area has generated many scholarly articles, not many books have been written on it. To avoid the redundancy of citing the works used for the discussion in this section, it suffices to state that what follows was delineated from the following works that are cited in the bibliography: Abraham (1996), Arnold (1992), Briggs (1997), Broer et al. (2003), Buffoni and Toland (2003), Crawford (1991), Demazure (2000), Dumortier et al. (1991), Françoise and Roussarie (1990), Gilmore (1993), Glendinning (1994), Postle (1980), Poston and Stewart (1998), Puu (2000), Sanns (2000), Saunders (1980), Schlomiuk (1993), Seeman (1977), Strogatz (2000), Thom (1989), Thompson and Michael (1982), Weisstein (n.d), Wiens (n.d), Woodcock et al. (1978), and Zhusubaliyev and Mosekilde (2003).

While mathematicians are generally consistent in defining the term "first order differential equation," this is not the case for the term "Bifurcation." Instead, they tend to view Bifurcation as a description of certain phenomena. A common perception is that a system undergoes a bifurcation if and only if the global behavior of the system, which depends on a parameter, changes when the parameter varies. Consequently, a general and a theoretical definition of Bifurcation can be delineated.

Bifurcation can be generally defined as a differential equation system that undergoes a qualitative change in its orbit structure, as one or more parameters of the dynamical system are changed. A Bifurcation is said to occur when a small smooth change made to the parameter values, or the Bifurcation parameters, of a system causes a sudden qualitative or topological change in its short-term dynamical behavior. Bifurcations result in both continuous systems, described as ordinary differential equations (ODEs), delay differential equations (DDEs), and partial differential equations (PDEs); and discrete systems, described by maps.

Theoretically, Bifurcations can be defined as the mathematical study of changes in the qualitative or topological structure of the integral curves of a vector field, or the solutions of a differential equation. Essentially, a Bifurcation is said to occur at a parameter value where a number of solutions changes. In essence, Bifurcation Theory examines structurally unstable dynamical systems. Dynamic stability refers to perturbations in the phase space—i.e. the stability of fixed points and limit cycles; and structural stability refers to perturbations in the function space—i.e. the topological stability of orbit structures.

Bifurcations have been typologized in two ways. One typology, which is less technical, identifies three types of Bifurcations: (1) Subtle Bifurcations, whereby an attractor changes type; (2) Catastrophic Bifurcations, whereby attractors appear out of, or disappear into, the blue: and (3) Explosive Bifurcations, whereby attractors drastically change size. The other typology, which is more technical, identifies two types of Bifurcations: (1) Local Bifurcations that take place through changes in the local stability properties of equilibria (i.e. fixed points), periodic orbits, or other invariant sets as parameters cross through critical thresholds; and (2) Global Bifurcations that take place when larger invariant sets of a system collide with one another, or with equilibria of a system. These Bifurcations cannot be detected purely by a stability analysis of the equilibria. Since the second typology is predominantly employed in Mathematics, a bit more discussion of their characteristics is in order.

A Local Bifurcation emerges when a parameter change causes the stability of an equilibrium to change. In continuous systems, this corresponds to the real part of an eigenvalue of an equilibrium passing through zero; in discrete systems, this corresponds to a fixed point having a Floquet Multiplier (relating to the class of solutions to linear differential equations) with modulus equal to one. In both cases, the equilibrium is non-hyperbolic at the Bifurcation Point (more on this later). The topological changes in the phase portrait of the system can be confined to arbitrarily small neighborhoods of the bifurcating fixed points by moving the Bifurcation parameter close to the Bifurcation Point—thus, the name "local."

Technically, the continuous dynamical system described by the ODE is considered as follows:

$$\dot{x} = f(x, \lambda) \quad f : \mathbb{R}^n \times \mathbb{R} \to \mathbb{R}^n.$$

A Local Bifurcation emerges at (x_0,λ_0), if the Jacobian Matrix $df_{x0,\lambda0}$ has an eigenvalue with zero real part. If the eigenvalue is equal to zero, the Bifurcation is said to be a steady state Bifurcation; if the eigenvalue is non-zero, but purely imaginary, then it is said to be a Hopf Bifurcation (more on this later).

For discrete dynamical systems, the following system is considered:

$$x_{n+1}=f(x_n,\lambda)$$

Consequently, a Local Bifurcation emerges at (x_0,λ_0), if the Jacobian Matrix $df_{x0,\lambda0}$ has an eigenvalue with modulus equal to one. If the eigenvalue is equal to one, the Bifurcation is either a Saddle-node—often called Fold Bifurcation when it appears in maps, Transcritical or Pitchfork Bifurcation. If the eigenvalue is equal to -1, it is a Period-doubling (or Flip) Bifurcation; otherwise, it is a Hopf Bifurcation.

The following are examples of Local Bifurcations:

(a) Hopf Bifurcation—a Bifurcation in which a fixed point of a dynamical system loses stability as a pair of complex conjugate eigenvalues of the linearization around the fixed point cross the imaginary axis of the complex plane.
(b) Horse Saddle Bifurcation—a saddle point that is a minimax: i.e. a local minimum of maximum depending on the intersecting plane used.
(c) Monkey Saddle Bifurcation—an example of an immersion, it is a surface defined by the equation to which it belongs to the class of saddle surface, and its name derives from the observation that a saddle for a monkey requires three depressions: two for the legs and one for the tail.
(d) Neimark (Secondary Hopf) Bifurcation—a Bifurcation through which a system loses its stable period one operation.
(e) Period-doubling (Flip) Bifurcation—a Bifurcation in which the system switches to a new behavior with twice the period of the original system.
(f) Pitchfork Bifurcation—a generic Bifurcation in which a symmetric solution changes its stability; it occurs generically in systems with symmetry.
(g) Saddle-node Bifurcation—a Bifurcation in which two fixed points of a dynamical system collide and annihilate each other.
(h) Transcritical Bifurcation—a Bifurcation characterized by an equilibrium having an eigenvalue whose real part passes through zero.

A Global Bifurcation emerges when "larger" invariant sets, such as periodic orbits, collide with equilibria. Such a collision leads to changes in the topology of the trajectories in the phase space which cannot be confined to a small neighborhood, as is the case with Local Bifurcations. The changes in topology extend out to an arbitrarily large distance—hence, the name "global."

The following are examples of Global Bifurcations:

(a) Blue Sky Catastrophe, when a limit cycle collides with a nonhyperbolic cycle.
(b) Global Saddle (Fold) Bifurcation, when a system is expressed in polar coordinates.
(c) Harmoclinic Bifurcation, when a limit cycle collides with a saddle point.
(d) Heteroclinic Bifurcation, when a limit cycle collides with two or more saddle points.
(e) Infinite-period Bifurcation, when a stable node and saddle point simultaneously occur on a limit cycle.

It should be noted that Global Bifurcations can also involve more complex sets such as chaotic attractors.

The codimension of a Bifurcation refers to the number of parameters which must be varied for the Bifurcation to emerge. This corresponds to the codimension of the parameter set for which the Bifurcation emerges within the full space of parameters. Saddle-node Bifurcations are the only generic Local Bifurcations which are really codimension-one; the others all have higher codimension. Nonetheless, Transcritical and Pitchfork Bifurcations are also often perceived to be codimension-one, because the normal forms can be written with only one parameter.

An example of a well-studied Codimension-two Bifurcation is the Bogdanov-Takens bifurcation, which is characterized by the fact that for the particular parameter value the vector field has a singularity whose linearized field has a double zero eigenvalue, while the other eigenvalues have nonzero real part. The latter condition is imperative.

The Bifurcation Diagram is a very useful tool employed to illustrate Bifurcations. It helps to show the possible long-term values—equilibria or periodic orbits—of a system function of a Bifurcation parameter in a dynamical system. It is normal practice to show stable solutions with a solid line and unstable solutions with a dotted line.

Named after their pioneer, Mitchell Feigenbaum, the Feigenbaum Constants are two mathematical constants used to express ratios in a Bifurcation Diagram. The first Fegenbaum Constant is expressed as follows:

$$\delta=4.669201609102990671853203821\ldots,$$

where sequence A006890 in the On-line Encyclopedia of Integer Sequences (OEIS) is the limiting ratio of each bifurcation interval to the next, or between the diameters of successive circles on the real axis of the Mandelbrot set. While Feigenbaum originally related this number to the Period-doubling Bifurcations in the logistic map, he later showed it to hold for all one-dimensional maps with a single quadratic maximum. As a result of this generality, every chaotic system that corresponds to this description will bifurcate at the same rate. This Feigenbaum's constant is often employed to predict when chaos will arise in such systems even before it does.

The second Feigenbaum Constant (sequence A006891 in the OEIS),

$$\alpha=2.502907875095892822283902873218\ldots$$

is the ratio between the width of a tine and the width of one of its two subtines (with the exception of the tine closest to the fold).

Both of the preceding numbers are applicable to a large class of dynamical systems. They are believed to be transcendental, albeit that claim remains to be proven.

Finally, Catastrophe Theory (which is also a special case of more general Singularity Theory in Geometry) considers the special case in the study of dynamical systems where the long-run stable equilibrium can be identified with the minimum of a smooth, well-defined potential function called Lyapunov Function. Since small changes in certain parameters of a nonlinear system can cause equilibria to appear or disappear, or to change from attracting to repelling and vice versa, this can lead to large and sudden changes of the behavior of a system. Investigated in a larger parameter space, catastrophe theory reveals that such Bifurcation points tend to occur as part of well-defined qualitative geometrical structures. Other aspects of Catastrophe Theory include Elementary Catastrophes, the potential functions of one active variable (Fold Catastrophe, Cusp Catastrophe, Swallowtail Catastrophe, and Butterfly Catastrophe), potential functions of two active variables (Hyperbolic Umbilic Catastrophe, Elliptic Umbilic Catastrophe, and Parabolic Umbilic Catastrope) and Arnold's notation, which are beyond the scope of this chapter. The interested reader will be well served by consulting the works mentioned earlier.

AFRICAN APPLICATIONS

Many works exist that have employed Bifurcations to examine various African phenomena in different academic disciplines, albeit only a handful of these works are discussed here. Before discussing these works, however, we will briefly talk about a phenomenon we had mentioned earlier: i.e. mathematical aspects of Bifurcations in ancient Egyptian artefacts. Indeed, classic Egyptians were quite involved with symbolism. Thus, their artefacts were designed and aligned cosmologically. And as discussions in earlier chapters have shown, the cosmological, astronomical, astrophysical, astrological and mathematical abilities of classic Egyptians were highly organized and quite advanced, and they performed truly remarkable feats of urban planning and architecture.

In their work on Auric Time Scale, Sergey Smelyakov, Geoff Stray and Jan Wicherink (2006a & 2006b) assert that in terms of proportions, the scale is reflected in the artefacts of Egypt. According to them, The Auric Time Scale Γ with the radix φ not only describes the crucial epochs in the known history of the Earth and humanity, it also adequately reflects the spectrum of the basic periods of Nature and society, including the geological and Solar activity cycles and artefacts of Egypt. In essence, embedded in Egyptian artefacts are representations of those events that have their impact on worldly events.

In order to get a sense of the connection between Egyptian artefacts and Bifurcations, it makes sense to quote verbatim what Smelyakov and his colleagues mean by Auric Time Scale. According to them, "In the narrow sense, the Auric Time Scale (ATS) is the series G of the Golden section powers (infinite to both ends), which is accompanied by the series G* = 2G of its double values; for its unity, phi^0 = 1 the average Solar cycle length T0 = 11.07 year, or Tropical year is taken" (Smelyakov et al., 2006a:1). They add that "In the broad sense, the ATS is a theory suggesting that for both principal aspects of time (periods and chronology), these series, T and T*, describe the length of the bulk of the basic periods in nature and society, and the major part of the most important historical and natural events in the evolutionary context, respectively" (Smelyakov et al., 2006a:1). I examine two examples of early Egyptian artefacts with aspects of Bifurcation in the following paragraphs: (1) ancient Egyptian calendars and (2) the pyramid texts of Set (Seth, Seti, Setekh, Setesh, Suty, Sutekh) of Nubet. A great deal has been written about these artefacts; thus, only brief descriptions are provided on them here.

According to Ronald Wells (1994 & 1997), ancient Egyptian calendars are important for investigating the development of ancient time measurement that guided classic Egyptians when to plant and when to harvest for successful food production. He points to the differences between the Delta and the Valley of the Nile in terms of the environment and the arable land available. He argues that these differences contributed to the development of the notion of duality which dominated early Egyptian thought and religion.

Wells observes that while religion in the north had its focus of solar imagery (for example, the pyramids at Giza and the sun tempes of Abu Ghurab), emphasis was placed mainly on stellar systems (for example, the temple of Satet at Elephatine) in the south. He notes that the earliest calendars used in Egypt appear to be lunar in origin and based on the moon's phases. These calendars were used to determine the dates of the religious year's festivals. In the south, the main festival was the "Going Forth of the Star Sirius" (Prt Spdt); in the north, the major festival was the "Birth of the Sun." Each of these festivals occurs approximately six lunar phases apart (Wells, 1994 & 1997).

Wells describes the depiction of the legend of the goddess Nut on the ceilings of a variety of temples and tombs (e.g., the tomb of Ramases VI in the Valley of the Kings) as holding the key to some of the astronomical observations which combined to form the Egyptian year. He notes that the Rameses VI ceiling has mirrored images of Nut—her body running back-to-back in a Siamese twin manner in an East-West formation along the center of the vaulted ceiling of the tomb, with her limbs outstretched and dropping down to the north and south walls. He points out that one twin represents the northern half of the sky in Egypt while the other represents the southern half. According to him, the stars on Nut's body represent the brightest stars of the Milky Way, the latter giving the impression of a diaphanous gown cloaking the body of a female. The stars of the constellation Cygnus, according to him, represents a mark of Bifurcation of the Milky Way into leg-like appendages, with the brightest star Deneb (Cyg) being located at the position of the birth canal. He adds that the Barque of Re traverses Nut's body as the sun passes through the heavens, and the ecliptic passes through the mouth of her head as formed by the Milky Way near the constellation of Gemini (Wells, 1994 & 1997).

The pyramid texts of Set—referred to by many other names such as god of darkness or evil, brother and enemy of Osiris, god of thunder and storm, the personification of evil in the battle against good, god of chaotic forces who commands both veneration and hostility—portray a complicated character. Probably a heraldic animal, Set is a quadruped with a gently curving muzzle, two appendages jutting from the top of its head and an erect tail remaining in a short Bifurcation. It appears on the mace head of King Scorpion at the end of the Predynasty era (http://www.egypt-topics.com).

According to Geoffrey Parrinder, the papyrus Jumilhac includes stories of Set that do not flinch from ascribing to him some very coarse and vulgar behavior. Two ideas are linked with the offerings of the sacred banquet. The first idea comprises pleasing gifts identified with the Eye of Horus. The second one comprises victims (relatives and followers of Horus and Osiris) slain by Set and his followers (Parrinder, 1985:137, 142).

The bifurcation of the struggle between Osiris and Set, which Ninian Smart suggests that in some ways prepared the ancient world for the acceptance of the Christian story, is as follows:

> Seth, brother of Osiris, had a beautifully shaped coffin made. He prepared a banquet and promised the coffin to the guest who would fit it. Only Osiris fitted it precisely, and as soon as he was lying inside Seth had the coffin fixed shut. He sealed the coffin with lead and had it thrown into the Nile. It eventually washed up on the other shore, where it was found by the distraught Isis, Orisis' wife. She had the body returned, but Seth got it and chopped it into pieces and scattered them throughout Egypt. Isis recovered all the bits—save the penis, for which she made a substitute—and reconstituted Osiris. Before embalming him she gave him new life—so much so that Osiris had intercourse with her, and their son, Horus, was thus conceived. Eventually, Horus fought and overcame Seth, but injured his eye in the process. His good eye came to be identified with the Sun and the bad eye with the Moon. The relationship to the Sun meshed in with the cult of the great god Re. Horus was often seen as a falcon, perhaps stretching his wings and invisibly, save to the mind's eye, hooding the whole sky (Smart, 1989:199).

The major lessons from the Bifurcations in the two examples are profitable even today. To this day, calendars are perceived to portend lucky/good and unlucky/evil days for people born on particular days. The civil calendar was standardized into a 365-day year with the division of the night and day into 12 hours each which was handed down to us through the Greeks and Romans.

In their article, "Genetic Evidence for a Paleolithic Human Population Expansion in Africa" (1998), David Reich and David Goldstein attempt to resolve the dates and magnitudes of dramatic expansions and size in human populations, besides the growth associated with agriculture that has been estimated by other scholars. Reich and Goldstein introduce two new statistical tests for population expansion (Within-Locus *k*-Test and Interlocus *g* Test), which employ variation at a number of unlinked genetic markers to investigate the demographic histories of natural populations. By analyzing genetic markers, Reich and Goldstein reveal highly significant evidence for a major human population expansion in Africa, but no evidence outside of Africa. They find that the inferred African expansion is estimated to have occurred between 49,000 and 640,000 years ago, certainly before the Neolithic expansion, and probably before the splitting of African and non-African populations. By showing a significant difference between African and non-African populations, Reich and Goldstein's analysis supports the postulate of a unique role of Africa in human evolutionary history, as has been suggested by most other genetic researchers. Also, Reich and Goldstein's findings support the "Out of Africa" model of modern human origins that the missing signal in non-African populations may be the result of a population bottleneck associated with the emergence of these populations from Africa (1998:8119).

The markers employed by Reich and Goldstein in their tests are "microsatellites," which were first identified in large-scale gene mapping projects but are now increasingly being used for inferring population parameters. These scholars note that microsattelites, which exhibit extensive "length" variations, are widely distributed throughout the genome, they seem to be selectively neutral, and they appear to conform reasonably well to a simple mutation process (stepwise mutation model), whereby mutations change the length by one or occasionally two units. Based on this mutation model, Reich and Goldstein develop the two statistical tests (mentioned in the preceding paragraph and to be discussed in the following paragraphs) to discern whether populations have been constant of growing in size (1998:8119).

For the Within-Locus *k*-Test for Population Expansion technique, Reich and Goldstein state that for a population of constant size genealogists tend to have a single ancient Bifurcation, which implies that most pairs of alleles are either closely or distantly related, with few in between. Thus, according to them, the distribution of allele lengths has discrete peaks that correlate with the descendants of each side of the ancient Bifurcation. They note that contrastingly, most of the Bifurcations tend to date back to the time of expansion: i.e. the genealogical tree is "comblike" and the allele length that results is more smoothly peaked (Reich and Goldstein, 1998:8119).

In order to differentiate between the ragged, multipeaked distribution expected for a constant population size and the smooth, single-peaked distribution expected for an expansion, Reich and Goldstein construct a statistic, denoted *k*, which is a decreasing function of the fourth central moment of a sample:

$$1/n\ \Sigma^{n}_{i=1}(X_i - \sim)^4$$

where:

n = number of chromosomes
$\sim$ = average allele length
X_i = individual allele lengths

They add that since the fourth central moment is related to the kurtosis, which increases the peakedness, the statistic *k* tends to decrease systematically with the degree of peakedness caused by expansion (Reich and Goldstein, 1998:8119).

Reich and Goldstein utilize computer simulations based on the coalescent algorithm of R. R. Hudson to set the parameters of the *k* statistic empirically. They trace genealogies backward in time from the sampled individuals to their most recent common ancestors and distribute stepwise mutations along the genealogies according to a random Poisson process. They then use the results of the simulations to set the parameters of the *k* statistic in order for the probability of a locus being positive when the population that is constant in size is constrained to a narrow range—between 0.515 and 0.55—for sample sizes greater than 10 and for a wide variety of population sizes and mutation rates. This computation yields the following statistic:

$$K = 2.5*\mathrm{Sig}^4 + 0.28*S^2 - 0.95/n - \mathrm{Gam}_4$$

Where:

S^2 = sample variance
Sig^4 = unbiased estimator for the variance squared
Gam_4 = unbiased estimator for the fourth central moment

Reich and Goldstein note that Sig^4 and Gam_4 are derived specifically for their analysis and that their validity is supported by computer simulation (1998:8119).

In implementing the approach, Reich and Goldstein conservatively set the probability of a positive k at 0.515 and employ a simple one-tailed binomial test to determine whether fewer loci were associated with a positive k than would be expected for a constant-sized population. They reason that since the expectation of k decreases with increasing kurtosis, such a reduction in the number of positive k values can be interpreted as a sign of population expansion (Reich and Goldstein, 1998:8120).

For the Interlocus g Test for Population Expansion technique, Reich and Goldstein focus on a feature of multilocus data sets that has no analog in studies of a single gene. Their rationale is that when populations are of constant size, the dates of the most ancient Bifurcations are subject to considerable variation from one locus to the other. They also reason that under conditions of growth, the most ancient Bifurcations tend to have similar dates at all loci. They note that the characteristic differences associated with demography will also be reflected in the variance of the variance of the allele length distributions, as a way for them to distinguish between the demographic scenarios. They assert that the variance of the variance is expected to be larger for constant-sized populations than for growing populations, specifically because the variance of an allele length distribution is contingent upon the ages of the few most ancient Bifurcations (Reich and Goldstein, 1998:8120).

In order to test for this statistical effect, Reich and Goldstein capitalize on the fact that there is an analytical expectation for the variance of the variance in a constant population:

$$4/3(E[V_j])^2 + 1/6E[V_j]$$

They substitute V, the average variance across the loci, for $E[V_j]$ to estimate the preceding quantity. They consider the ratio, g, of the observed value to the expected value in order to formulate the test explicitly. They then take a sufficiently low value of g as a sign of population expansion, noting that a useful and interesting feature of this ratio is that its expectation and confidence intervals are essentially independent of mutation rate times population size ($N\mu$) and nearly independent of sample size (Reich and Goldstein, 1998:8120).

In regards to the Paleolithic human population expansion in Africa, Reich and Goldstein state that one tetranucleotide and one dinucleotide microsattelite data set, each of 30 unlinked loci, provide information about genetic diversity represented in hundreds of individuals and several populations around the world. In the tetranucleotide data, the "within locus" k test shows that only two populations—the San and the Sotho-Tswana, both in Africa—give a significant signal of expansion, both with P values of < 0.01. The interlocus g test applied to the data also reveals a difference between African and non-African populations, as the four lowest *g* values are in Africa and clearly lower than those found elsewhere in the world (Reich and Goldstein, 1998:8120).

In the dinucleotide data, Reich and Goldstein find the within-locus *k* test to produce no significant *P* values, possibly because as they demonstrate with computer simulations the test loses power for higher values of the mutation rate. Nonetheless, with the interlocus *g* test, Reich and Goldstein show that the North-Central African population shows a significant sign of expansion, with $P < 0.037$. The significance of the detected expansion increases even further to $P < 0.006$ when these scholars drop an exceptionally variable locus (D13S122), which has a variance in the worldwide sample of 89.2 compared with a range of 1.0 to 17.2 for the other 29 loci. Contrastingly, the non-African populations do not show signs of expansion when the high-variance locus (D13S122) is dropped, albeit the *g* values for these populations are all below 1, suggesting expansions (Reich and Goldstein, 1998:8121).

To estimate the date of the expansion, Reich and Goldstein define an expansion time, in addition to its associated pre-expansion population size and factor of expansion, to be "allowable" if computer simulations using these three parameters generate 90 percent confidence intervals that include the observed values of *g* and the average variance across loci. Using *N* as the pre-expansion population size, the researchers consider 60 values of $N\mu$ between 0.05 and 25, 60 values of the expansion time from 0 to $10/N$ generations in the past, and factors of sudden expansion ranging from 3 to 100. Applying this procedure to the data, they calculate allowed dates utilizing 29 of the dinucleotide loci typed in the North-Central African population, ignoring the anomalously high-variance locus, and incorporating the estimated variation in the mutation rate, which is $0.28\,\mu^2$. With an average dinucleotide mutation rate estimated at 5.3×10^{-4} per generation, and a generation estimate of 25 years, they make the following inferences (Reich and Goldstein, 1998:8122–8123):

> The maximum pre-expansion population size for the North-Central African population is 6,600, the lower bound for the post-expansion population size is 8,400, and the allowed dates are between 49,000 and 640,000 years ago—certainly predating the advent of agriculture. Crude estimates of the maximum likelihood surface for the date, based on computer simulations, indicate that the distribution is binomial, and thus that a point estimate may not be very informative. The positions of the peaks for the various factors of expansion, however, constrain the maximum likelihood estimate between 148,000 and 364,000 years, consistent with the expansion having occurred around or before the split of modern human populations in Africa [estimated to have occurred 75,000–287,000 years ago using the dinucleotide data, and dated to similar times using other data].

Reich and Goldstein conclude by noting that the method of allowed dates seems to be robust because it produced similar ranges of dates for widely varying expansion factors. They also caution, however, that the real pattern of growth is likely to have been considerably more complicated, involving repeated periods of expansion and possibly even contractions, and that it is not clear how these complications would affect their inferences (Reich and Goldstein, 1998:8123).

Ernest Wilson and Kelvin Wong in their article titled "African information Revolution: A Balance Sheet" (2003) provide a policy and institutional framework to describe and analyze the diffusion of information technology and the global information revolution in Africa and the major factors that influence the diffusion. They begin by examining regional diffusion and find substantial cross-national diffusion differences across the continent, with considerable variation in regional diffusion in telephone, Internet, radio, and television. They note that this pattern undermines technologic and economic explanations as the only determinants of variation in diffusion. They then conduct an analysis of the information revolution in Sub-Sahara Africa based on a policy framework entailing four key policy balances: (1) public and private initiatives, (2) monopoly and competition "markets," (3) domestic and foreign ownership or control, and (4) the diffusion of the information revolution. They find that a necessary condition for an explanation of the diffusion of the information revolution hinges upon a policy and institutional framework that incorporates these four policy balances (Wilson and Wong, 2003:155).

As Wilson and Wong point out, Africa, like other regions in the Third World, stands at the entrance of a global information revolution that is pregnant with many opportunities. New technologies like the Internet and cellular telephones are everywhere, as do traditional media like radio. In every country, the local press provides information and commentaries on the latest information and communications gadgets. Despite all this, the breadth and complexity inherent in this technology make it difficult to reach unambiguous conclusions about the extent and meaning of the information revolution for Africa. In light of this situation, Wilson and Wong observe that it is important to provide a coherent overview of these critical changes because Africa's future growth and the wellbeing of its people will hinge in part on the continent's capacity to make these new resources widely available (Wilson and Wong, 2003:155–156).

Bifurcations in Wilson and Wong's study can be found in their discussion of Phase I: 1980-1990 of Africa's progress toward information and communications technology (ICT) reforms. According to these scholars, the shifting balances in ICT among African countries neither happened at once nor did they happen the same way in all countries. Instead, these shifting balances, with occasional reversals and surprises, over the course of more than 20 years, between 1980 and 2000, emerged slowly and fitfully. The efforts were led by reformers in both the public and private sectors (Wilson and Wong, 2003:166).

Wilson and Wong point out that in the 1980s, African ICT sectors were organized around the classic balances of the *ancient regime*. With slight variations, ICT policy was quite stable and for the most part reflected the interests of the states' elites that held power, who preferred the public, monopolistic, domestic and centralized system. The system operated well for some years after political independence, with the respective ICT Ministries making virtually all tariff, investment, pricing and other decisions within the centralized bureaucracies. The states owned, operated and regulated all communications functions, from post to telephone and telegraph (Wilson and Wong, 2003:166).

According to Wilson and Wong, while the systems appeared stable over many years and widely accepted by elites worldwide, their demise actually started in the late 1970s and 1980s. As early as 1967, the prelude to reform began when the Ethiopian government bifurcated its Post, Telephone and Telegraph (PTT) Office monopoly into separate offices: one responsible for postal functions, and the other in charge of telecommunications operations. This Bifurcation of traditional PTTs into post and telecommunications, two distinct and separate organizations, Wilson and Wong argue, is imperative for decentralizing the old PTTs. Approximately 12 years later in 1979, the PTTs in Burundi and Lesotho followed Ethiopia's lead. Nonetheless, as these scholars show through a time-series pictogram, it was not until the early 1980s that many PTTs in Africa took the first steps toward bifurcating their operations (Wilson and Wong, 2003:166).

As Wilson and Wong also observe, an important and exceptional instance of institutional Bifurcation that reflects policy change took place in 1981 when Senegal's international telephone operations separated from the PTT. Named TeleSenegal, the new parastatal was launched to operate the international telephone network, while the PTT, the Office of Post and Telecommunications (OPT), maintained and operated the national telephone network. The launching of TeleSenegal as a parastatal was significant because it allowed it to gain "autonomy of decision and management" and was able to access its own funding without the same bureaucratic hurdles that confronted OPT. The reform, however, was short lived, when in 1985 the OPT and TeleSenegal were officially disbanded as postal and telecommunications functions were bifurcated (Wilson and Wong, 2003:167).

Sonatel was launched in 1985 to replace TeleSenegal and the telecommunications portion of OPT. The new entity was charged with managing both national and international telephone networks. While it was officially "freed from direct governmental control and supervision," a number of institutional ties with the Ministry of Communication continued. In essence, some "minimal conditions of liberalization" had been implemented that served as the basis for the reforms that followed. The combination of Bifurcation and the introduction of limited autonomy, Wilson and Wong suggest, made Sonatel a different kind of institution (2003:167).

In his article tilted "Formation of Bifurcating Chromitite Layers of the UG1 in the Bushviel Igneous Complex, an Analogy with Sand Volcanoes" (2004), P. A. M. Nex describes the Bushveld Igneous Complex in South Africa as the world's largest layered intrusion and asserts that the complex is justly famed for its magnetic ore deposits of chromite, platinum group elements, and vanadium. They note that the complex is made up of volcanic rocks (Rooiberg Group), a mafic layered suite (Rustenburg Layered Suite), and sheeted granites (Lebowa Granite Suite) emplaced onto and within sediments of the Transvaal Supergroup at c. 2054–2060 Ma. Nex points out that the layered mafic rocks of the complex have been divided into the following zones based on their cumulate lithologies: (a) the Lower and Marginal Zones are made up of laterally discontinuous norites; (b) the Critical Zone is comprised of varying cycles of chromitite, pyroxemite, norite and anorthosite, which together form spectacularly layered rocks; (c) the Main Zone is dominated by gabbronorites and minor anorthosites; (d) the Upper Zone is made up of gabbros and ferrogabbros, and is noted for its magnetic layers. He states that chromitite layers can only be found in the Critical Zone, which is subdivided into upper and lower portions, with the boundary between the two portions marked by the first appearance of cumulus plagioclace. He adds that the focus of his study is on the UG1 chromitite layer, which is the lowest of the Upper Group chromite layers and the first substantial chromitite layer within the upper Critical Zone and is underlain by a much thicker anorthosite than any other chromite layer (Nex, 2004).

According to Nex, any model employed to analyze chromitite Bifurcations must explain the fundamental correlation between symmetry of Bifurcations in terms of the domal structures. He mentions that previous analysts have not fully addressed the question of the origin of Bifurcations before, and the occurrence and significance of domes have not been well investigated. He notes that disrupted chromitite or anorthosite layering shows one of either (a) ductile deformation of chromitite, or (b) ductile deformation of chromitite and anorthosite together, or (c) brittle deformation of chromitite, and that no evidence has been shown for brittle deformation of anorthosis on its own. He therefore suggests that anorthosite was capable of flow after consolidation of chromitite layers (Nex, 2004).

A model that achieves these demands, according to Nex, must compare the structures in the UG1 chromitite layer and footwall with structures found in sedimentary rocks as a result of fluidization, particularly sand volcanoes and sand boils. He states that an investigation of cross-sections of both sand volcanoes and UG1 domes reveals that the features in both have similar geometrical relationships and, possibly, a similar method of formation (Nex, 2004).

Nex goes on to point out that some scholars have noted the formation of sand boils or volcanoes in sedimentary rocks in preserved sedimentary structures in ancient rocks and also forming in currently unconsolidated sediments. He observes that such formations occur when a layer of unconsolidated sediment is fluidized and transport of fluid and suspended grains takes place through overlying sediment to the sediment-water interface. He notes that the slurry of fluid plus grains is then "emptied" at the sediment-water interface, forming domal structures that are related to volcanoes. He also states that similarly, fashion fluidization of the UG1 footwall is suggested, for which there is plenty of evidence, including the formation of dome structures, the geometrical relationship between Bifurcations and domes, and the occurrence of chromitite layers on a variety of scales (Nex, 2004).

Finally, Nex envisages that while the footwall anorthositic was still unconsolidated possibly to a depth in excess of 10m, liquefaction of part of this interval occurred, and at a similar time chromite commenced accumulating. He points out that chromite accumulation provided the "background sedimentation" when periodic extrusion of plagioclase crystals plus magna occurred at the magna-cumulate-pile interface. He states that the plagioclase crystals would have formed sub-circular lenses of anorthosite that punctured background chromite accumulation. He adds that the thickness of the chromitite layers forming during periodic extrusion of anorthosite and magna would thus depend on the presence or absence of anorthosite "volcanoes" and the rates at which both plagioclase and chromite crystals were accumulating. He concludes that this mechanism explains primary accumulation features, including the formation of dome structures, the geometrical relationship between Bifurcations and domes, and the occurrence of chromite layers on a variety of scales (Nex, 2004).

Mohamed Mostafa in his article, "Factors Affecting Organisational Creativity and Innovativeness in Egyptian Business Organisations: An Empirical Investigation" (2005), seeks to further the understanding of how managers in Egypt perceive creativity and innovativeness. He also investigates the construct validity of two measures of creativity barriers in order to gain further insights into the factors that stimulate or hinder creativity in Egypt.

For his research, Mostafa employed a sample of 170 managers. He states that the sample slightly favored high-level managers due to the managers' functional areas. Research subjects completed a 77-item survey instrument designed to assess stimulants-to-creativity in the organizational workplace. In addition, he also used a 17-item barriers-to-creativity survey instrument to determine barriers to creativity in Egypt's organizations (Mostafa, 2005).

Mostafa finds a statistically significant difference in attitude towards organizational creativity based on the managers' functional areas in their organizations. He also detects that the greater the education of a manager, the more likely s/he will adopt creative and innovative activities. The t-test procedure shows no generation gap in the managers' attitudes towards organizational creativity. The results also show that male managers have significantly favorable attitudes towards creativity compared to their female counterparts (Mostafa, 2005).

Bifurcations in Mostafa's study are evident in rigid rules, which are found to be inversely related to creativity and innovation. Citing Chaos Theory, Mostafa notes that it suggests that change is constant, its consequences unforeseen and not subject to control or accurate prediction. According to Mostafa, this situation is evident in two statements in the survey questionnaire that were loaded into the study's fourth factor (as in Factor Analysis technique), with reliability alpha 0.71; the overall mean is 3.12. The two statements deal with fear of failure and regulations to follow, respectively. These statements emerged prominent because, in the words of Mostafa, "Management has a tendency to preserve the established traditions, and therefore many rules and standard procedures were set for the employees to follow and keep them under control" (Mostafa, 2005).

Finally, due to the preceding results, Mostafa states that because a certain amount of instability is essential for change, Egyptian managers should stay flexible and be willing to manage creativity without rigid rules. He believes that at certain times, things in organizations may seem quiet and calm; yet, at other times, an iterative process suddenly takes off, leading to a Bifurcation or branching phenomenon that allows a system to go in an entirely new direction (Mostafa, 2005).

In their article titled "Future Ecosystem Services in a Southern African River Basin: A Scenario Planning Approach to Uncertainty" (2006), Erin Bohensky, Belinda Reyers and Albert Van Jaarsveld employ scenarios in one of the sub-global components of the Millennium Ecosystem Assessment (a four-year initiative to explore relationships between ecosystem services and human wellbeing at multiple scale) to investigate four possible futures in a Southern African river basin—specifically, the Gariep Basin. According to these scholars, due to its ability to show spatial and temporal dynamics, the scenario exercise reveal major trade-offs in ecosystem services in space and time and the essence of a multiple-scale scenario design. They note that at sub-global scales, scenarios are powerful tools for communication and engagement of decision makers, especially when designed to generate solutions to specific problems. Thus, they argue that scenario planning has the potential to be a critical instrument in conservation, as many calls are being made to help define and achieve sustainable visions for the future (Bohensky et al, 2006:1051).

Bohenski and her colleagues experiment with several approaches to describe the implications of the scenario Bifurcations for ecosystem services. First, they use an integrated dynamic system model to generate results. Second, they employ an interactive approach and ask users to draw arrows to indicate direction and magnitude of change in ecosystem services and human wellbeing under each scenario relative to the existing condition. Third, they utilize spider diagrams to illustrate trends in the narratives. They expect the scenarios to manifest differently within the brain and, thus, define four zones based on biophysical and socioeconomic characteristics: "(1) urban areas, notably Gauteng Province, which depend to a large degree on ecosystem services from other regions; (2) the 'grain basket,' the agriculturally productive grasslands and water-rich highlands; (3) the densely populated, largely rural, and poor Great Fish River; and (4) the 'arid west,' a low-rainfall, sparsely populated, mostly rural expanse of land where many mining operations are concentrated" (Bohenski et al., 2006:1056).

There are several major findings in the work of Bohenski and her colleagues that are particularly noteworthy for the current study. First, the direction and magnitude of change in ecosystem services in each scenario and region expected by users' of the Gariep Basin are described as (a) a sharp increase, +2; (b) a slight increase, +1; (c) no change, 0; (d) a slight decrease, -1; or (e) a sharp decrease, -2. The investigators distinguish between provisioning services such as food in which an increase signifies higher levels of service production and regulating and supporting services such as biodiversity in which an increase translates to an improvement in the condition of the service. Although freshwater provides both types of services, the investigators focus on its regulating services to be in sync with the expanded definition of water resources under the 1998 South African Water Act (Bohenski et al., 2006:1058).

Second, of major conservation concern are trade-offs and synergies between ecosystem services and biodiversity. A strong economic argument for conservation can be made for the maintenance of some services such as nature-based tourism, medicinal plants, and crop pollination that has a clear link to biodiversity. A fundamental link can also be made between biodiversity and human wellbeing, since it allows the rural poor to maintain diverse livelihoods based on ecosystem services (Bohenski et al., 2006:1058–1059).

Third, a contradiction exists between the local resources scenario and the "tragedy of the commons," suggesting that some ecosystem degradation can be avoided even in the absence of strong central government control through self-regulating local institutions. Nonetheless, this scenario also suggests that basin-scale measures are imperative to protect downstream water resources from upstream impacts; and in the adapting mosaic scenario, global interventions are needed to govern the global commons. As people begin to comprehend the links among ecosystem service, biodiversity, and human wellbeing, with the attendant coordination at multiple scales reflecting this understanding, then the policy reform scenario, like technogarden, will work (Bohenski et al., 2006:1059).

Jonathan Sherratt and Gabriel Lord in their article, "Nonlinear Dynamics and Pattern Bifurcations in a Model for Vegetation Stripes in Semi-arid Environments" (2006), present a detailed examination of a model for vegetation stripes based on competition for water by using the patterned solutions in the full nonlinear model and employing Numerical Bifurcation Analysis of both the pattern ODEs (ordinary differential equations) and the model PDEs (partial differential equations). They demonstrate that patterns exist for a wide range of rainfall levels and, in particular, for much lower rainfall than have been considered by previous researchers. They also demonstrate that for many rainfall levels, patterns with a variety of different wavelengths are stable, with mode selection dependent on initial conditions. This situation, they believe, raises the possibility of hysteresis that allows them to utilize solutions of the model to show that pattern selection hinges upon rainfall history in a relatively simple way.

Sherratt and Lord begin by noting that vegetation in many semi-arid environments is not homogenous, but rather is self-organized in spatial patterns. They point out that on flat ground, these patterns appear randomly; on gentle slopes, a striped pattern is the norm, with bands of vegetation up to 250 m wide, separated by gaps of up to 1 km, running along the contours. These patterns can be found in a wide range of grasses and small shrubs and are hard to detect on the ground. They note that the patterns were first observed in aerial photographs of Sub-Saharan Africa in the 1950s (Sherratt and Lord, 2006:1).

The main concern of Sherratt and Lord is with C. A. Clausmeier's 1999 study in which he synthesized ideas in earlier computer-based simulation models to develop a couple of differential equations for vegetation $U(x, t)$ and surface water $W(x, t)$. The equations represent the basic processes of plant growth in proportion to water availability, plant loss, and plant dispersal. Sherratt and Lord state that water input is the result of rainfall and water is lost through a combination of evaporation and active uptake by plants. They postulate that the last term is nonlinear in the plant density U, since the presence of plant roots in the soil increases water infiltration. They also note that the spatial coordinate X runs uphill, so that there is an active flow of water in the negative X direction (Sheratt and Lord, 2006:1–2).

Sherratt and Lord observe that the focus of the Klausmeier model is on water flow downhill and, thus, cannot be used to predict patterning on flat ground. They point out, however, that M. Rietkerk and his colleagues and J. van de Koppel and his co-researchers in a number of their works have subsequently proposed a number of models that incorporate an additional variable, as well as plant density and surface water. Rietkerk and his colleagues added water within the soil as their additional variable and performed a detailed numerical Bifurcation study on a major variant of this model to show that bifurcations leading to patterning are subcritical. Koppel and his co-researchers added herbivores as an additional variable in their study. Since these scholars had added an additional variable in their works, Sherratt and Lord therefore argue that these works are more realistic extensions of that of Klausmeier (Sherratt and Lord, 2006:2).

Nonetheless, Sherratt and Lord are quick to point out that the Klausmeier model and its extensions are not the only theoretical explanations for vegetation stripes. They point out that Lejeune and his colleagues have also employed a detailed model that combines short-range activation and long-range inhibition between neighboring plants. Sherratt and Lord note that in the Lejeune et al. model, activation is due to shading of one plant by another, with the difference in length scales of the processes due to the root system within the soil being much more extensive than the parts of the plants above ground. In their model, Lejeune and his colleagues use slope as a selector as opposed to an initiator of spatial patterning. Sherratt and Lord also note that E. Meron and his colleagues used yet another method, with a model formulated in terms of plant density and water in the soil, in which the latter has a transport term based on porous media theory. Sherratt and Lord further point out that recently, Meron's group proposed a related model with separate variables for surface water and soil water. Again, as Sherratt and Lord sum up, all of these models predict pattern formation on flat as well as sloping ground (2006:2).

Also, for Sherratt and Lord, the Klausmeier, Rietkerk-van de Koppel, Lejeune and Meron models all provide plausible explanations for vegetation patterning. Thus, a full understanding of patterning in the various models is essential for distinguishing the different mechanisms, a task Sherratt and Lord suggest their study undertakes (2006:2).

Accordingly, Sherratt and Lord begin by using linear analysis to study the low amplitude patterns. Next, they present numerical results to which the analytical approximations are compared. After this, they show from numerics that patterned solutions exist for significantly smaller rainfall levels than previously considered. Their approach involves applying the Bifurcation package AUTO to the ordinary differential equations governing patterned solutions. The pattern they study moves uphill, and the speed of this movement is entered as an additional parameter in their calculations (Sherratt and Lord, 2006:3).

Sherratt and Lord's model yields a number of major findings. First, wavelength emerges as the most accessible property of vegetation stripes in the field. It needs to be observed only a single time and it can be done though aerial photography, avoiding logistical difficulties and expensive ground-based observations. Consequently, the essential prediction of theoretical models for vegetation stripes is pattern wavelength and its variation with model parameters (Sherratt and Lord, 2006:9–10).

Second, unlike earlier models that made predictions on pattern wavelength based on the most unstable wavenumber, Sherratt and Lord's results reveal that it is wrong to make the simple assumption that the observed pattern has the most unstable wavenumber. Instead, pattern selection is nonlinear and history-dependent. Specifically, Sherratt and Lord focus on patterning in response to changes in rainfall, albeit analogous results are tenable for changes in grazing levels (2006:10).

Third, vegetation stripes can be initiated either by the destabilization of the homogenous vegetation state or by the introduction of plants to desert regions. Unlike the Klausmeier model that predicts a critical rainfall level at which pattern formation occurs (linear stability analysis enables calculation of this critical level and also the corresponding spatial wavelength), Sherratt and Lord's results reveal that as rainfall is further decreased, pattern wavelength remains the same, although the form of the pattern changes gradually, with the vegetation stripes becoming thinner and the stripe separation increasing. According to Sherratt and Lord, the pattern wavelength only changes at a rainfall level that is significantly less than that at which patterns first arise by a factor of 0.7 for the specific parameter set. Comparatively, the most unstable wavelength changes by about 30 percent during this change in rainfall. Once pattern wavelength increases, the longer wavelengths persist even when rainfall increases again (Sherratt and Lord, 2006:10).

Fourth, on the one hand, patterns that might arise through the introduction of plants to desert areas cannot be investigated by employing the stability and Bifurcation approach because the desert steady state is always linearly stable in the Klausmeier model, so that even with high rainfall it needs a large perturbation to get vegetation growth. Sherratt and Lord's model, on the other hand, shows that the lowest rainfall levels allow patterns so that there is a single vegetation stripe on the domain, no matter its size (Sherratt and Lord, 2006:10).

In light of the preceding findings, Sherratt and Lord conclude that their results show that pattern selection from the Klausmeier model differs significantly from the "most unstable mode" rule suggested by earlier investigators. Moreover, Sherratt and Lord add, their own work makes it possible to predict pattern wavelength based on parameter values and on the basic knowledge of rainfall history (2006:11).

Finally, in their article, "The Dynamics of Measles in Sub-Saharan Africa" (2008), Matthew Ferrari and his eight colleagues investigate why measles epidemics in Niger, particularly in the capital Niamey, are highly episodic. They note that while vaccination has almost eliminated measles in most parts of the world, the disease remains a major killer in some high birth rate countries of the Sahel region, of which Niger is a part. Their models demonstrate that this variability arises from powerful seasonality in transmission, generating high amplitude epidemics, within the chaotic domain of deterministic dynamics. They observe that in practice, this situation leads to frequent stochastic fadeouts, interspersed with irregular, large epidemics. Their metapopulation model shows that increased vaccine coverage, but still below the local elimination threshold, could lead to increasingly variable major outbreaks in highly seasonally forced contexts. Such erratic dynamics, they add, emphasize the essence both of control strategies that address build-up of susceptible individuals and efforts to mitigate the impact of large outbreaks when they occur.

According to Ferrari and his coauthors, among the acute infections, the epidemic dynamics of measles are the best understood. Forced mainly by seasonal variations in infection, powerful herd immunity leads to a tendency for multi-annual outbreaks. The deep inter-epidemic troughs that result can cause local stochastic extinction of infection in towns below a critical community size of 300,000–500,000. This highlights the epidemiological impact of spatial heterogeneity in host distribution that can also influence complex spatiotemporal epidemic patterns. Another factor that can have a strong impact on epidemic dynamics has to do with demographic heterogeneities in the recruitment of susceptible individuals. In theory, strong seasonal forcing can push chaotic dynamic in the measles attractor. In practice, however, measles dynamics and persistence are more consistent with weaker seasonality that is driven by epidemic cycles shaped by demographic heterogeneities in space and time (Ferrari et al., 2008).

For Ferrari and his fellow investigators, Niger is an important case to understand the dynamics and control of vaccine-preventable childhood infections in a high birth rate nation where these infections remain a major problem. Niger is located in the western Sahel and ranges from several densely populated cities in the south to desert climates in the north that is sparsely populated by nomadic pastoralists. With a population of about 13 million, Niger's birth rate is among the highest in the world at 50.73 births per 1,000 people. Ferrari and company point out that routine single-dose measles vaccine distribution through the Expanded Program on Immunization (EPI) was initiated in 1987, but it was not until 2004 that Niger's first measles-only Supplementary Immunization Activities (SIAs) targeted all children aged nine months to 14 years, achieving an estimated coverage rate of 99 percent of the target population. Ferrari and his colleagues' analysis of the temporal dynamics and spatial synchrony of measles outbreaks at the local scale, however, shows that the appearance of regular, annual outbreaks is an artefact of averaging erratic and asynchronous local epidemics (Ferrari et al., 2005).

To parameterize their model, Ferrari and his colleagues focus on the relatively well-documented time series of incidence from 1986 to 2002 from Niamey, a city of almost 750,000 inhabitants. Empirical patterns over the last 30 years show highly erratic measles outbreaks, with monthly case reports from 1986 to 2004 revealing occasional large outbreaks followed by years of very few cases. In a similar fashion, annual measles incidence rates in Niamey between 1975 and 1985 ranged from one to five percent, consistent with this irregular pattern (Ferrari et al., 2005).

Ferrari and his colleagues note that measles epidemics in Niamey decline at the onset of the rainy season, no matter the magnitude of the outbreak. This shows that powerful seasonal forcing of transmission may be driving irregular, fragile dynamics even in such a large, high birth rate population. Ferrari and his fellow scholars investigate this issue employing a stochastic time series Susceptible-Infected-Removed (TSIR) epidemiological modeling framework, which has been applied successfully to measles dynamics elsewhere by other researchers. The TSIR model allows Ferrari et al. to estimate the form of seasonality in transition. First, they use sinusoidal forcing to illustrate the general dynamical consequences of varying seasonal amplitude, which shows a Bifurcation Diagram for a simple deterministic TSIR model with a fixed 14-day infectious period and sinusoidal forcing in transmission rate as a function of seasonal amplitude and birth rate. Seasonal transmission is modeled as a cosine wave:

> B (t) = (mean β) (1 + αcos (2nt). X a axis, the amplitude of the seasonal forcing, α; y axis, annual birth rate per 1,000 people in the population (Ferrari et al., 2005).

In addition, at low seasonal amplitude, seasonality = 0.2, the dynamic resembles historical patterns in the industrialized world (e.g., London): a dynamic transition from annual to biennial cycles as birth rate declines from high levels. Contrastingly, at high seasonal amplitude, seasonality = 0.6, the range of birth rates at which the system exhibits stable 1–4-year cycles decreases and the dynamics become chaotic over a broad range of birth rates (Ferrari et al., 2005).

Also, Ferrari and his colleagues estimate seasonal variation in the transmission rate in Niamey explicitly by applying the TSIR model to 17 years of monthly data from the city. They use a Bayesian state space approach to account for uncertainty in the reporting rate. Their estimated seasonality in the transmission rate reveals a single peak, roughly in antiphase to the seasonal rainfall profile. They state that a possible mechanistic explanation for this pattern hinges on the increase in urban density in the dry season due to seasonal migration from outlying agricultural areas (Ferrari et al., 2005).

Furthermore, Ferrari and his colleagues estimate the seasonal variation in transmission rate by fitting a TSIR model with imperfect binomial reporting to the 17-year-long time series of monthly incidence in Niamey (1986–2002) using Bayesian Markov chain Monte Carlo methods. They specify the unobserved time series of measles cases as a TSIR model:

$$I_{t+1} \sim \mathrm{NB}(\beta_m S_t I_t^{\alpha}, I_t)$$

where:

NB(a, b) = a negative binomial process
S = the number of susceptible hosts
I = the number of infested hosts
B_m = the month-specific transmission rate
α = a tuning parameter to account for non-linearities in transmission

Ferrari and his co-researchers then take a time step as 0.5 months in order for the TSIR model to be coupled to a binomial observation model in which the observed number of cases each month is distributed as binomial:

$$(I_{t-1} + I_t, P_{obs})$$

where:

P_{obs} = the reporting probability for cases

They finally evaluate the effect of vaccination on outbreak dynamics using a metapopulation model made up of 40 local communities representing 39 arrondissements plus Niamey. They model coupling among patches as a power function of distance that is parameterized on the basis of the 2001–2005 spatial resolved data. They also work on the assumption that the routine vaccination targeted young children across the entire metapopulation.

CONCLUSION

There is an old Chinese curse that says "May you live in interesting times" (Laszlo, 1991). Indeed, that we do live in interesting times, as futurist and systems philosopher Erwin Lazlo writes in his appropriately titled book, *The Bifurcation Paradigm: Understanding the Changing World* (1991), is hardly a matter of dispute. Even more interesting is the fact that a phenomenon that was present in artefacts of ancient Egyptians is playing a significant role in the current revolution in the field of Mathematics: i.e. the Chaos Revolution. Indeed, Bifurcation Theory has helped to shift the explicit mathematical analytic emphasis from computation of individual evolutions (solutions) to considering the whole phase space as it contains all possible evolutions. The interest has moved away from individual systems and towards the study of dynamical properties that are persistent in small perturbations of a given system. Parallel to this development is the introduction of parameters into dynamical systems in order to comprehend deformations or the unfolding of these deformations in a systematic manner.

In sum, it can be stated that the emphasis in dynamical systems shifted from the explicit analytic computation of a given evolution to the consideration of generic properties of families of dynamical systems deforming a given one, where the methods have become more geometric and qualitative. By combining geometric and algebraic methods, the quantitative element also resurfaces, as the geometric descriptions can be traced in detail to the original physical model equations.

Chapter Twelve

Fractals

INTRODUCTION

The major focus of this chapter is on the evidence of the existence of fractals in African designs. But before we delve into such a discussion, it makes sense to begin with a brief discussion of the fractal process itself and the connection between chaos and African fractals.

The fractal process is a graphical approach whereby the regions are subdivided until the sub-regions are less than one screen pixel in size. In essence, the strength of a fractal process is that no matter how small a region gets, one can still generate more detail simply by evaluating the procedure on the new, smaller region.

Thus, the fractal process is a special case of procedural process in that a fractal object generally has to be *self-similar*: that is, as one generates finer and finer object detail, that detail begins to look like the object itself. A procedural process is that instead of defining and storing a large data set of points to describe a shape, a procedure for computing the value (color, intensity, and shape) at any given point, given previously computed values for surrounding points, is defined. A good example pointed out by Christopher Watkins and Larry Sharp (1992:86) is a leaf. The branching patterns of a leaf look very similar to the shape of the leaf itself. A careful examination of the branches with a magnifying glass reveals that the branches have finer branches, which in turn have finer branches, and so on.

FRACTALS IN AFRICAN DESIGNS

The following discussion draws greatly from the findings of Eglash (1997b, 1999), who has done the most work in this area, and Bangura's work titled *Chaos Theory and African Fractals* (2000). As Eglash (1999) points out, the most mathematically significant aspect of doubling in African religion occurs in the divination ("fortunetelling") techniques of vodum and its religious relatives. The Ifa system, for example, is characteristic of what mathematicians would call "stochastic," that is, it operates by pure chance. It is based on tossing pairs of flat shells or seeds split into two. Each lands open-side or closed-side. They are connected by a doubled chain to make four pairs. Each group of four pairs gives one of the 16 divination symbols that tell the future of the diviner's client.

Nonetheless, a closely related divination system called Cadena has a non-stochastic element, or what mathematicians would refer to as "deterministic chaos." Cadena is based on a symbolic code in which each symbol, represented by a set of four vertical dashed lines drawn in the sand, stand for some archetypical concept (desire, travel, health, etc.) with which narratives about the future are assembled.

The Bamana divination, like the Cantor set (pathological curves), is based on recursion. Bamana divination begins with four horizontal dashed lines, drawn rapidly, so that there is some random variation in the number of dashes in each. The dashes are then connected in pairs, such that each of the four lines is left with either one dash or none. The narrative symbol is then constructed as a column of four vertical marks, with double vertical lines representing an even number dashes and single lines representing an odd number. There are two possible marks in four positions, thus 16 possible symbols. The random symbol production is repeated four times.

Both tarumbeta and owari mathematical game boards' marching-group dynamics are governed by triangular numbers. What is special about these types of cellular automata boards is the underlying concept of recursion—the way in which a kind of mathematical feedback loop can generate new structures in space and new dynamics in time.

Legba, god of chaos or "god of crossroads," is represented by a fork because the answer could be either yes or no—one never knows which path he will take. For divination, in which a "path" (question) is often pushed for further questions, the image becomes one of endless bifurcation. The bifurcating uncertainties of Legba are like a positive feedback loop, amplifying deviation and noise (i.e. random variation).

In the iconic carvings of the Baule, a contrast between a negative feedback loop, to create stability, and the positive feedback of uncontrolled disorder exists. For example, a carving of two caimans (relatives of the alligator) bit-ing each other's tail depicts the chief and the people in balance—if one bites, the other will bite back. Another example is a carving showing that power creates the appetite for more power—little fish are eaten by bigger fish, which then become even bigger fish.

A form of music indigenous to Nigeria has something like a white noise (i.e. random noise with the highest algorithmic complexity) distribution of sounds. For example, the random music of the Birom is a flute ensemble designed to allow each musician to express individual feelings through whatever idiosyncratic noise (or even silence) s/he chooses. This results in an indeterminate process in which the sounds produced by the players are not obstructed by the conscious attempt to organize the rhythms and harmonies.

In Yoruba folktale, one learns about the trickster Eshu as the "lord of random." There is a coupling between the orderly work of Olirun and the unpredictable spirit of Eshu, similar to the negative feedback/positive feedback mentioned earlier.

From this relatively small number of examples, it is quite obvious that African Fractal Geometry is not a singular body of knowledge, but rather a pattern of resemblance that can be seen when a variety of African mathematical ideas and practices is described. And, indeed, it is not the only pattern possible. In short, it makes sense to see African fractals as just another moment in a historical sequence.

Elaborate cornrow braids on an African woman's head can be viewed as more than an affinity with culture or a fashion statement. The intricate patterns are also useful for learning about African fractals—geometric patterns that are repeated on smaller and smaller scales to produce intricate designs that are beyond the scope of classical or Euclidean Geometry, which refers to the "ordinary" system of geometry based on the parallel postulate—i.e. a basic assumption that states whether through a point P, not on a given line l, there exists none, one, or more than one line parallel to the given line l.

Fractal geometry has emerged as one of the most exciting frontiers in the fusion between mathematics and information technology. Fractals can be observed in many of the swirling patterns produced by computer graphics, and they have become a vital tool for modeling in the natural sciences. While fractal geometry can allow one to get into the far reaches of high tech science, its patterns are surprisingly common in traditional African designs. Also, some of the basic concepts in fractal geometry are fundamental to African knowledge systems: quantitative techniques, symbolic systems, engineering, architecture, games, traditional hairstyling, textiles, sculpture, painting, carving, metalwork, and religion.

Eglash demonstrates the principle with a tool on his Web site (http://www.cohums.ohio-state.edu/comp/eglash.dir/afractal.htm) that allows one to create computer-generated graphics simulating the fractal branching patterns found in braided hairstyles. Eglash's book, *African Fractals: Modern Computing and Indigenous Design*, explores how fractals are evident in all levels of African society, including architecture and art, politics, and religion.

As Eglash explains, although most people learn Euclidean Geometry in school, few study fractal geometry (because it is difficult), which plays a significant role in the computer modeling processes in the hard sciences. Meanwhile, Eglash adds, fractal geometry has long been a theme in Africa, with a wide variety of local cultural associations, including kinship, labor practices, politics, and religion.

Eglash's research began in the 1980s while investigating settlement architecture in Central and West Africa. Aerial photographs of various settlement compounds revealed that many were composed of circular structures enclosed in other circles, or rectangles within rectangles, and that the compounds were likely to have street patterns in which broad avenues branched into very small footpaths.

As Eglash notes, at first he thought it was just from unconscious social dynamics. But during his field work, he adds, he found that fractal designs also appear in a wide variety of intentional designs—carving, hairstyling, metalwork, painting, textiles—and the recursive process of fractal algorithms are even employed in African quantitative systems. Eglash adds that in the design rationales and cultural semantics of many African geometric figures, as well as in indigenous quantitative systems (additive progression, doubling sequences, binary recursion) and symbolic systems (ionic symbols for feedback loops, equiangular spirals, infinity), there are abstract ideas and formal structures that closely parallel some of the fundamental aspects of fractal geometry. These results, he concludes, are congruent with recent developments in complex systems theory, which suggest that pre-modern, non-state societies were neither utterly anarchic, nor frozen in static order, but rather utilized an adaptive flexibility that capitalized on the nonlinear aspects of ecological dynamics.

While in Africa, Eglash encountered some of the most complex fractal systems that exist in religious activities, such as the sequence of symbols used in sand divination, a method of fortune telling found in Senegal. Some of his other findings include the use of sophisticated mathematical ideas in everyday objects. In the arid region of the Sahel, artisans produce windscreens by utilizing a scaling design that gives them the maximum effect—keeping out the wind-driven dust—for the minimum amount of effort and material.

A number of advancements are taking place in the study of African fractals. The prominent ones are discussed in the following paragraphs. What the following discussion shows is how modern computational tools are being used for mathematical experimentation and theorizing about African fractals.

The African Fractals in Development project is about a boot-strapping approach to investigating indigenous African science for education and development. The principal investigator of the project is Ron Eglash. The co-principal investigators are Christian Sina Diatta and Egondu Onyejekwe. The former is the director of the Institut de Technologie Nucleaire Aplique at Universite Cheikh Anta Diop of Dakar, Senegal. He is originally from the lower Casamance, and he coordinates links between the rural sites and university activities. The latter is a computer scientist at Ohio State University, and he manages links between Diatta's operations at the Universite and the support activities in the United States through the African Women Global Network (this organization's Web site is http://www.osu.edu/org/awognet/). In addition, Nfally Badiane, a graduate student in anthropology at the Universite Cheikh Anta Diop of Dakar is in charge of the Ziguinchor operations.

The project began with Eglash's visual observation that aerial photos of traditional African settlements tend to have a fractal structure (scaling in street branching, recursive rectangular enclosures, circles of circular dwellings, etc.). This was quantitatively confirmed in Eglash and Broadwell (1989), where they applied a 2-dimensional Fourier transform (an analysis of frequencies that may be embedded in a body of data) to digitized photo images to estimate the fractal dimension from the slope of the spectral density function. Subsequent study by Eglash (including a year of field work under the Fulbright program in Central and West Africa) revealed that these architectural fractals result from intentional designs, not simply unconscious social dynamics, and that recursive scaling structures can be found in other areas of African material culture.

The major objective of the investigators is to test the possibility that indigenous knowledge can be employed in a boot-strapping approach to development. The investigators conceptualize "boot-strapping" as a development approach in which one begins with indigenous knowledge under local control, and self-installs modern technological abilities.

The project focuses on the creation of an indigenous science center in the city of Ziguinchor, in the lower Casamance region of Senegal. The choice of this site hinged on the excellent examples of fractal Ethnomathematics (mathematical practices of identifiable groups) in the area (Eglash, Diatta and Badiane 1994). The goals of the center are as follows: (1) utilize indigenous knowledge for the benefit of the people and natural environment of the region; (2) create such benefits by combining indigenous knowledge with modern scientific frameworks and technological capabilities; (3) minimize the need for external support and maximize the potential for expansion; (4) create a model for other indigenous science centers in other regions, with the expectation that a network of these institutions could promote inter-ethnic cooperation and political stability.

In terms of methodology, while the indigenous center itself is used to define local problems, the investigators began their work with some obvious challenges. To begin with, increasing salinization of the area has led to a decreasing quality of life. Next, external economic pressures (e.g., the move to cash-crop production for export and tourism) and the migration to cities have disrupted many of the beneficial attributes of traditional life in the area (e.g., gender equality), increased disease (e.g., AIDS), and damaged the environment. The highly decentralized social structure of the Djola and the long historical conflicts with the north render policy recommendations particularly ineffective. Thus, the investigators believe that by encouraging the development of indigenous knowledge in modern technological frameworks, the Djola will be able to carry out their own empirical studies, educational projects and technological production to meet the challenges. Thus, the project's methodology focuses on the following six aspects:

1. *Architecture for an indigenous science.* The physical location of the center embodies the philosophy of the project, employing local materials and labor, and utilizing indigenous design practices. For instance, the French non-governmental organization (NGO) ENDA has devised a modern water collecting system based on the traditional Djola impluvia. At the same time, it makes use of recycled materials and generates solar electricity and heating. It is believed that these factors will ensure that the center will not shut down due to lack of utilities once the initial funding period is completed, while creating an infrastructure for its research facilities, school room, visitors' center, and educational office.
2. *An indigenous knowledge database.* A difficulty with indigenous knowledge research noted by the investigators is the separation between projects and institutions. They point out that, for example, traditional agricultural practices have been studied in the Casamance region by projects under Senegalese, American, French, Italian and Arabic governments, in addition to those of NGOs. Ethnobotany, ethnomedicine, and ethnomathematics of the local culture is not under-researched, but rather research findings are so widely diffused that this information is not readily available. The investigators believe that the advent of computer technology will remedy this problem. By linking a multimedia database at the science center in Ziguinchor with a similar facility at the University of Dakar, investigators hope to synthesize past research together with contemporary studies, ac-

cess university resources, and facilitate Internet communication with the global research community.

3. *Agro-ecology*. Since botanical knowledge is quite extensive in traditional Djola culture, the investigators believe that this suggests an obvious application to recent developments in combining ecological biodiversity with agricultural practices and nutritional concerns. They cite the example of gastropod consumption being increasingly common due to the salinity increase. They also suggest that it could be possible to identify which species could best be harvested, or even cultivated, and thus encourage use of indigenous species while improving local food supplies and decreasing the motivation of forest destruction. They add that, most importantly, such innovations could be derived experimentally by the local population, which could increase the chances of them being widely disseminated and utilized. And they note that long-term records of the experiments and their results would be maintained in the database.
4. *Technological expertise*. The extreme degree of the recycling of artificial materials is one of the notable characteristics of contemporary African life. However, while this is true of construction and craft materials, it is not so of electronics. By providing an electronics shop, the investigators foresee the science center as being able to recycle cast-off electronic items, build up a parts supply, and even produce its own instrumentation. One example they note is that of salinity which can be monitored from widespread sites throughout the Casamance, if simple instrumentation is widely available. Oscilloscopes could then be created from cast-off television parts, meaning that similar electronic shops could be set up elsewhere. Junk dealers even sell computer parts nowadays.

   A side benefit from this expertise is that technological syncretism could be made available to the arts. For now, the growth of discos means only Euro-American audio technology and light shows, but traditional African artists could also take advantage of electronic media as well, acting as a force for cultural survival in the face of the Western cultural encroachment.
5. *Science education*. For the investigators, the utilization of traditional knowledge in math, science and engineering education is perhaps the most important function of the center. They cite the example of the Djola who are excellent musicians and have developed a "whistle language" as well as a drum communication system that allows them to communicate over long distances. According to the investigators, these attributes make the Djola vocabulary rich in analogs for oscillation, frequency, phase, and other elements of signal processing. Indeed, this is an excellent basis for physics education. The investigators plan to develop an introduction to

basic physics, using amplified sound from Ekonting visually displayed on an oscilloscope, to demonstrate how the modern physics concepts are translated by the Djola indigenous knowledge system. They made a request to the Tektronix Foundation to donate an oscilloscope for the task. Portland's Black Educational Center also expressed an interest in participating in an exchange program to further develop the physics lab for encouraging more African American students to become interested in math and science. The investigators also point out that since it is widely acknowledged that some version of the Ekonting is the precursor of the American banjo, the connection to America heritage, both African and European, is quite profound.

For the investigators, here again, the major aspect for developing such a synthesis is to facilitate local empowerment. Students will be poised to use indigenous knowledge to learn ecology, medical sciences, and physical sciences in ways that encourage both the continuation of Djola tradition and prepare them for an increasingly technological world.

6. *Financial stability*. The investigators stress the need for independence from external funding to ensure the future of the project. They prefer self-generated income. They note that, for example, the Casamance has a large number of annual visitors, but presently there is no effort to create what is called "green tourism," despite an intense interest in both the natural and the cultural facets of the area. The investigators believe that such efforts would easily dovetail with the educational activities of the center.

When Eglash returned from Africa, one of his colleagues advised him to focus on scaling patterns in African hairstyles. An enthusiastic group of students at Evergreen State University volunteered their programming skills to help create a multimedia lesson on African fractals. The Hairstyle Storyboard Web site that has been developed utilizes a style referred to as "the braids of threads," from Yaoundé, Cameroon, to explicate branching fractals (www.cohuns.ohio-state.edu/comp/eglash.dir/afmulti.htm).

The "fractal hairstyles" module guides users, step by step, through the creation of a three-dimensional fractal, beginning with the initial design and then mathematically determining the ratio of each iteration. The major goal of the principal investigators of this project (Ron Eglash, Gloria Glimer, T. Q. Berg, and Jaron Sampson) is to combine the images, software and video on African fractals.

One can learn about African fractals from a video project titled "Math and Science through the Eyes of Culture." Produced by the Nebraska Telecommunication Network, the 28 minutes and 45 seconds video is distributed by GPN (acronym not spelled out). Narrated by the host, Julie Valentine, the video focuses on the works of three brilliant, contemporary scholars. The first is Tom Hull, a lecturer in mathematics who specializes in calculus and analytical geometry, who demonstrates how origami (arts and science of folding paper; the word is Japanese in origin, but the activity itself dates back thousands of years) can be used to explain some of the intricate aspects of geometry. The second is Eloy Rodriguez, a professor of environmental science at Cornel University, who shows how herbs from Africa and Latin America can be utilized in modern medical practices. And the third is Eglash who demonstrates the existence of fractals in African designs. The video can be ordered by postal address: 1800 North 33rd Street (68583), P. O. Box 80669, Lincoln, Nebraska 68501-0669; by telephone: 1.800.228.4630 or 1.402.472.2007; by fax 1.800.306.2330; or by e-mail: gpn@unlinfo.unl.edu.

There is a Web-based program that allows an individual to simulate African fractals. The Web site is www.osu.edu/org/awognet/research/Fractals/DevAfrican/Dev-African.html. The following fractal exercises are available:

1. Braided Hair
2. Cairo
3. Ethiopian Cross
4. Fulani Blanket
5. Logone-Birni
6. Mokoulek

Developed for the Macintosh computer and designed to accompany Eglash's book (1999), the FractaSketch African Edition software is an interactive color tool for designing and drawing intricate African fractals. Users create fractals from basic shapes that they enter graphically. In addition, the program comes with many ready-to-use fractal images, from natural-looking objects to order and chaos. The program provides infinite drawing levels, proportional line thickness capabilities, and the ability to compute fractal dimensions and save images in either "Compact Format," Postscript (TM) or in PICT file format. Some of the program's expansions include the following: color transition tools; window linking—a tool to allow fractal shapes created in one window to link with fractal shapes created in others; support of multiple frame generation—great for animation movies; extensive editing and object control; and high resolution printing output.

The producer of the software program, which comes in either disk version or as an e-mail attachment, is Bernt Wahl. It can be ordered through postal mail: Dynamic Software, P. O. Box 13991, Berkeley, California 94701; by telephone: 510.644.0139; or by e-mail: wahl.org/dynamic.

The study of African fractals has also found a home at the International Study Group on Ethnomathematics (ISGEm). This organization was founded in 1985 by math educators Gloria Gilmer, Umbiratan D'Ambrosio, and Rick Scott. Since then, the organization has sponsored programs and business meetings at the annual conferences of the National Council of Teachers of Mathematics (NCTM-USA) and at the International Congress of Mathematics Education (ICME). In 1990, ISGEm became an affiliate of NCTM-USA. ISGEm publishes two newsletters each year.

D'Ambrosio coined the term Ethnomathematics to describe the mathematical practices of identifiable groups—for example, African Fractals. Ethnomathematics is sometimes used specifically for small-scale indigenous societies; but in its broadest sense, the "ethno" prefix can refer to any group, be it a national society, a labor community, a religious tradition, a professional class, etc. The focus is therefore on the culture in which mathematics arises. While mathematics is often associated with the study of "universals," it is nevertheless important to recognize that often something that is thought of as being universal is merely universal to those who share a culture and a historical perspective.

Thus, the mathematical practices that are being emphasized by ISGEm include not only formal symbolic systems, but also spatial designs, practical construction techniques, calculation methods, measurement in time and space, specific methods of reasoning and inferring, and other cognitive and material activities. ISGEm thrives on the fact that there is now abundant empirical evidence to support the notion that people in all societies devise their own ways of doing mathematics, no matter their technological level or what they may have learned in school. The organization therefore strives to increase people's understanding of the cultural diversity of mathematical practices, and to make this knowledge applicable to education and development. (More information about ISGEm can be found at its Web site: http://www.cohums.ohio-state. edu/comp/isgem.htm.)

Work on African fractals has also found space in *Virtual Dimension*, an anti-manifesto of the so-called digital revolution, which critically examines the connections between emerging technologies, architecture, production, critical theory, the body and society's current obsession with cyberspace. The objective of this forum is to facilitate an open arena/site from which to project an interdisciplinary event of all that is digital, mediated and prosthetic. In essence, no manuscript exists. Rather, it is an assemblage of loosely disseminated fragments, critiques and tactics which are infinitely porous, and thus represent a series of de-stratified architectural/cultural positions, thresholds and fields.

Essays, interviews, QuickTimeVR, video, hypertext and digital images are incorporated throughout the project in order to create ruptures, or fissures, equivalent to the streams of consciousness in the Joycian sense (this is akin to literalist James Joyce's line in *Finnegan's Wake*, "there quarks of Muster Mark, when he postulated their existence). The purpose of employing the multimedia platform of printed text and CD-ROM is to make explicit and literalize the modalities, ideas and processes inherent in the digital and virtual media themselves. It is through this fusion of fuzzy aggregates that multiplicities and the "will to virtuality," the absence of presence, can best be understood. (Additional information can be retrieved through the project's Web site: http://www.papress.com/virtual/index.html.)

CONCLUSION

As Eglash observes, it is not mere coincidence that Benoit B. Mandelbrot, who is dubbed the father of fractals, was influenced by the work of Georg Cantor, who studied in Egypt; and that Cantor, in turn, was influenced by the work of Fibonacci, who was educated in Tunisia. In addition, Mandelbrot reported that he was inspired by H. E. Hurst's work on flood variations over several centuries and concluded that it could be characterized in terms of a scaling component. H. E. Hurst, it turns out, lived for 62 years in Egypt and was able to demonstrate long-term scaling in Nile flood records because of the accurate "nilometer" readings going back fifteen centuries. One would not know all this by reading Eurocentric texts on Fractals. In the words of the great African scholar Carter G. Woodson,

> From the teaching of science the Negro was likewise eliminated. The beginnings of science in various parts of the Orient were mentioned, but the Africans' early advancement in this field was omitted. Students were not told that ancient Africans of the interior knew sufficient science to concoct poisons for arrowheads, to mix durable colors for paintings, to extract metals from nature and refine them for development in the industrial arts. Very little was said about the chemistry in the method of Egyptian embalming which was the product of the mixed breeds of Northern Africa, now known in the modern world as "colored people" (1933:18–19).

Indeed, learning about African Fractals does contribute towards rendering obsolete these myths and misconceptions.

Chapter Thirteen

African-centered Automated Generation of Metadata

INTRODUCTION

The capabilities of generating and collecting data, observes Faleh Alshameri (2006), have been increasing rapidly. The computerization of many business and government transactions with the attendant advances in data collection tools, he adds, has provided huge amounts of data. Millions of databases have been employed in business management, government administration, scientific and engineering management, and many other applications. This explosive growth in data and databases has generated an urgent need for new techniques and tools that can intelligently and automatically transform the processed data into useful information and knowledge (Chen et al. 1996). This chapter explores the nature of data mining and how it can be used in doing research on African issues.

Data mining is the task of discovering interesting patterns from large amounts of data where the data can be stored in databases, data warehouses, or other information repositories. It is a young interdisciplinary field, drawing upon such areas as database systems, data warehousing, statistics, machine learning, data visualization, information retrieval, and high-performance computing. Other contributing areas include neural networks, pattern recognition, spatial data analysis, image databases, signal processing, and many application fields, such as business, economics, and bioinformatics.

Data mining denotes a process of nontrivial extraction of implicit, previously unknown and potentially useful information (such as knowledge rules, constraints, regularities) from data in databases. The information and knowledge gained can be used for applications ranging from business management, production control, and market analysis to engineering design and scientific exploration.

There are also many other concepts, appearing in some literature, carrying a similar or slightly different definitions, such as knowledge mining from databases, knowledge extraction, data archaeology, data dredging, data analysis, etc. By knowledge discovery in databases, interesting knowledge, regularities, or high-level information can be extracted from the relevant sets of data in databases and be investigated from different angles, thereby serving as rich and reliable sources for knowledge generation and verification. Mining information and knowledge from large databases has been recognized by many researchers as a key research topic in database systems and machine learning and by many industrial companies as an important area with an opportunity for major revenue generation. The discovered knowledge can be applied to information management, query processing, decision making, process control, and many other applications. Researchers in many different fields, including database systems, knowledge-based systems, artificial intelligence, machine learning, knowledge acquisition, statistics, spatial databases, and data visualization have shown great interest in data mining. Furthermore, several emerging applications in information providing services, such as on-line services and the World Wide Web, also call for various data mining techniques to better understand user behavior in order to ameliorate the service provided and to increase business opportunities.

MINING MASSIVE DATA SETS

Recent years have witnessed an explosion in the amount of digitally-stored data, the rate at which data are being generated, and the diversity of disciplines relying upon the availability of stored data. Massive data sets are increasingly important in a wide range of applications, including observational sciences, product marketing, and the monitoring and operations of large systems. Massive datasets are collected routinely in a variety of settings in astrophysics, particle physics, genetic sequencing, geographical information systems, weather prediction, medical applications, telecommunications, sensors, government databases, and credit card transactions. The nature of these data is not limited to a few esoteric fields, but, arguably, to the entire gamut of human intellectual pursuits, ranging from images on Web pages to exabytes ($\sim 10^{18}$ bytes) of astronomical data from sky surveys (Hambrusch et al. 2003).

There is a wide range of problems and application domains in science and engineering that can benefit from data mining. In several of these fields, techniques similar to data mining have been used for many years, albeit under different names (Kamath 2001). For example, in the area of remote sensing, rivers and boundaries of cities have been identified using image understanding methods. Much of the use of data mining techniques in the past has been for data obtained from observations of experiments, as one-dimensional signals or two-dimensional images. These techniques, however, are increasingly attracting the attention of scientists involved in simulating complex phenomena on massively parallel computers. They realize that, among other benefits, the semi-automated approach of data mining can complement visualization in the analysis of massive datasets produced by the simulations.

There are different areas which provide for the use of data mining. The following are some examples:

(a) *Astronomy and Astrophysics* have long used data mining techniques such as statistics that aid in the careful interpretation of observations that are an integral part of astronomy. The data being collected from astronomical surveys are now being measured in terabytes ($\sim 10^{12}$ bytes), because of the new technology of the telescopes and detectors. These datasets can be easily stored and analyzed by high performance computers. Astronomy data present several unique challenges. For example, there is frequently noise in the data due to the sensors used for collecting data: atmospheric disturbances, etc. The data may also be corrupted by missing values or invalid measurements. In the case of images, identifying an object within an image may be non-trivial, as natural objects are frequently complex and image processing techniques based on the identification of edges or lines are inapplicable. Furthermore, the raw data, which are in high-dimensional space, must be transformed into a lower-dimensional feature space, resulting in a high pre-processing cost. The volumes of data are also large, further exacerbating the problem. All these characteristics, in addition to the lack of ground truth, make astronomy a challenging field for the practice of data mining (Grossman et al. 2001).

(b) *Biology, Chemistry, and Medicine*—informatics, chemical informatics and medicine are all areas where data mining techniques have been used for a while and are increasingly gaining acceptance. In bioinformatics, which is a bridge between biology and information technology, the focus is on the computational analysis of gene sequences (Cannataro et al. 2004). Here, the data can be gene sequences, expressions, or protein information. Expressions mean information on how the different parts of a sequence are activated, whereas protein data represent the biochemical and biophysical structures of the molecules. Research in bioinformatics related to sequencing of the human genome evolved from analyzing the effects of individual genes to a more integrated view that examines whole ensembles of genes as they interact to form a living human being. In medicine, image mining is used on the analysis of images from mammograms, MRI scans, ultrasounds, DNA micro-arrays and X-rays for tasks such as identifying tumors, retrieving images with similar characteristics, detecting changes, and genomics. In addition to these tasks, data mining can be employed in the analysis of medical records.

In the chemical sciences, the information overload problem is becoming staggering as well, with the chemical abstract service adding about 700,000 new compounds to its database each year. Chemistry data are usually obtained either by experimentation or by computer simulation. The need for effective and efficient data analysis techniques is also being driven by the relatively new field of combinatorial chemistry, which essentially involves reactivating a set of starting chemicals in all possible combinations, thereby producing large datasets. Data mining is being used to analyze chemical datasets for molecular patterns and to identify systematic relationships between various chemical compounds.

(c) *Earth Sciences, Climate Modeling, and Remote Sensing* are replete with data mining opportunities. They cover a broad range of topics, including climate modeling and analysis, atmospheric sciences, geographical information systems, and remote sensing. As in the case of astronomy, this is another area in which the vast volumes of data have resulted in the use of semi-automated techniques for data analysis. Earth science data can be particularly challenging from a practical view point, and they come in many different formats, scales, and resolutions. Extensive work is required to pre-process the data, including image processing, feature extraction, and feature selection. It suffices to say that the volumes of earth sciences data are typically enormous, with the NASA Earth Observing System expected to generate over 11,000 terabytes of data upon completion. Much of these data is stored in flat files, not databases.

(d) *Computer Vision and Robotics* are characterized by a substantial overlap. There are several ways in which the two fields can benefit each other. For example, computer vision applications can benefit from the accurate machine learning algorithms developed in data mining, while the extensive work done in image analysis and fuzzy logic for computer vision and robotics can be used in data mining as well, especially for applications involving images (Kamath 2001). The applications of data mining methodologies in computer vision and robotics are quite diverse. They include automated inspection in industries for tasks such as detecting errors in semiconductor masks and identifying faulty widgets in assembly line productions; face recognition and tracking of eyes, gestures, and lip movements for problems such as lip-reading; automated television studios, video conferencing and surveillance, medical imaging during surgery as well as for diagnostic purposes, and vision for robot motion control. One of the key characteristics of the problems in computer vision and robotics is that they must be done in real time (Kamath 2001). In addition, the data collection and analysis can be tailored to the task being performed as the objects of interest are likely to be similar to one another.

(e) *Engineering*—with sensors and computers becoming ubiquitous and powerful, and engineering problems becoming more complex, there is a greater focus on gaining a better understanding of these problems through experiments and simulations. As a result, large amounts of data are being generated, providing an ideal opportunity for the use of data mining techniques in areas such as structural mechanics, computational fluid dynamics, material science, and the semi-conductor industry. Data from sensors are being used to address a variety of problems, including detection of land mines, identification of damage in aerodynamic systems (e.g., helicopters) or physical structures (e.g., bridges), and nondestructive evaluation in manufacturing quality control, to name just a few. In computer simulation, which is increasingly seen as the third mode of science, complementing theory and experiment, the techniques from data mining are yet to gain widespread acceptance (Marusic et al. 2001). Data mining techniques are also employed in studying the identification of coherent structures in turbulence.

(f) *Financial Data Analysis*—most banks and other financial institutions offer a wide variety of banking services such as checking, savings, and business and individual customer transactions. Added to that are credit services like business mortgages and investment services such as mutual funds. Some also offer insurance and stock investment services. Financial data collected in the banking and financial industries are often relatively complete, reliable and of high quality, which facilitate systematic data analysis and data mining. Classification and clustering methods can be used for customer group identification and targeted marketing. For example, customers with similar behaviors regarding banking and loan payments may be grouped together by multidimensional clustering techniques (Han et al. 2001). Effective clustering and collaborative filtering methods such as decision trees and nearest neighbor classification can help in identifying customer groups, associate new customers with an appropriate customer group, and facilitate targeted marketing. Data mining can also be used to detect money laundering and other financial crimes by integrating information from multiple databases, as long as they are potentially related to the study.

(g) *Security and Surveillance* comprise another active area for data mining methodologies. They include applications such as fingerprint and retinal identification, human face recognition, and character recognition in order to identify people and their signatures for access, law enforcement or surveillance purposes. Data mining techniques can also be used in tasks such as automated target recognition.

The preceding areas have benefited from the scientific and engineering advances in data mining. Added to these are various technological areas that produce enormous amounts of data, such as high energy physics data from particle physics experiments that are likely to exceed a petabyte ($\sim 10^{15}$ bytes) per year and data from the instrumentation of computer programs run on massively parallel machines that are too voluminous to be analyzed manually. What is becoming clear, however, is that the data analysis problems in science and engineering are getting more complex and more pervasive, giving rise to a great opportunity for the application of data mining methodologies. Some of these opportunities are discussed in the following subsection.

REQUIREMENTS AND CHALLENGES OF MINING MASSIVE DATA SETS

In order to conduct effective data mining, one needs to first examine what kind of features an applied knowledge discovery system is expected to have and what kind of challenges one may face at the development of data mining techniques. The following are some of the challenges:

(a) *Handling of different types of high-dimensional data.* Since there are many kinds of data and databases used in different applications, one may expect that a knowledge discovery system should be able to perform effective data mining on different kinds of data. Massive databases contain complex data types, such as structured data and complex data objects, hypertext and multimedia data, spatial and temporal data, remote sequencing, transaction data, legacy data, etc. These data are typically high-dimensional, with attributes numbering from a few hundreds to the thousands. There is an urgent demand for new techniques for data retrieval and representation, new probabilistic and statistical models for high-dimensional indexing, and database querying methods. A powerful system should be able to perform effective data mining on such complex types of data as well.

(b) *Efficiency and scalability of data mining algorithms.* With the increasing size of data, there is a growing appreciation for algorithms that are scalable. To effectively extract information from a huge amount of data in databases, the knowledge discovery algorithms must be efficient and scalable to large databases. Scalability refers to the ability to use additional resources such as the central processing unit (CPU) and memory in an efficient manner to solve increasingly larger problems. It describes how the computational requirements of an algorithm grow with problem size.

(c) *Usefulness, certainty and expressiveness of data mining results*. Scientific data, especially data from observations and experiments, are noisy. Removing the noise from data, without affecting the signal, is a challenging problem in massive datasets. Noise, missing or invalid data, and exceptional data should be handled elegantly in data mining systems. The discovered knowledge should accurately portray the contents of the database and be useful for certain applications.

(d) *Building reliable and accurate models and expressing the results*. Different kinds of knowledge can be discovered from a large amount of data. These discovered kinds of knowledge can be examined from different views and presented in different forms. This requires the researcher to build a model that reflects the characteristics of the observed data and to express both the data mining requests and the discovered knowledge in high-level languages or graphical user interfaces, so that the discovered knowledge can be understandable and directly usable.

(e) *Mining distributed data*. The widely available local and wide-area computer networks, including the Internet, connect many sources of data and form huge distributed heterogeneous databases, such as the text data that are distributed across various Web servers or astronomy data that are distributed as part of a virtual observatory. On the one hand, mining knowledge from different sources of formatted or unformatted data with diverse data semantics poses new challenges to data mining. On the other hand, data mining may help disclose the high-level data regularities in heterogeneous databases which can hardly be discovered by simple query systems. Moreover, the huge size of the database, the wide distribution of data, and the computational complexity of some data mining methods motivate the development of parallel and distributed data mining algorithms.

(f) *Protection of privacy and data security*. When data can be viewed from many different angles and at different abstraction levels, it can threaten the goal of ensuring data security and guarding against the invasion of privacy (Chen et al. 1996). It is important to study when knowledge discovered may lead to an invasion of privacy and what security measures can be developed to prevent the disclosure of sensitive information.

(g) *Size and type of data*. Science datasets range from moderate to massive, with the largest being measured in terabytes. As more complex simulations are performed and observations over long periods at higher resolution are conducted, the data will grow to the petabyte range. Data mining infrastructure should support the rapidly increasing data volume and the variety of data formats that are used in the scientific domain.

(h) *Data visualization.* The complexity and noise of massive data affect data visualization. Scientific data are collected from variant sources by using different sensors. Data visualization is needed to use all available data to enhance an analysis. Unfortunately, a difficult problem may emerge when data are collected on different resolutions, using different wavelengths, under different conditions, with different sensors (Kamath 2001). Collaborations between computer scientists and statisticians are yielding statistical concepts and modeling strategies to facilitate data exploration and visualization. For example, recent work in multivariate data analysis involves ranking multidimensional observations based on their relative importance for information extraction and modeling, thereby contributing to the visualization of high dimensional objects such as cell gene expression, profile, and image.

MINING SPATIAL DATABASES

The study and development of data mining algorithms for spatial databases are motivated by the large amount of data collected through remote sensing, medical equipment, and other instruments. Managing and analyzing spatial data became an important issue due to the growth of the applications that deal with geo-reference data. A spatial database stores a large amount of space-related data, such as maps, pre-processed remote sensing and medical imaging data. Spatial databases have many features distinguishing them from relational databases. They carry topological and/or distance information, usually organized by sophisticated, multidimensional spatial indexing structures that are accessed by spatial data access methods and often require spatial reasoning, geometric computation, and spatial knowledge representation techniques. Another difference is the query language that is employed to access spatial data. The complexity of the spatial data type is another important feature (Palacio et al. 2003).

The explosive growth in data and databases used in business management, government administration and scientific data analysis has created the need for tools that can automatically transform the processed data into useful information and knowledge. Spatial data mining is a subfield of data mining that deals with the extraction of implicit knowledge, spatial relationships, and other interesting patterns not explicitly stored in spatial databases (Koperski et al. 1995). Such mining demands an integration of data mining with spatial database technologies. It can be used for understanding spatial data, discovering spatial relationships, relationships between spatial and non-spatial data, constructing spatial knowledge databases, reorganizing spatial databases, and optimizing spatial queries. It is expected to have wide applications in geographic imaging, navigation, traffic control, environmental studies, and many other areas where spatial data are employed (Han et al. 2001).

A crucial challenge to spatial data mining is the exploration of efficient spatial data mining techniques due to the huge amount of spatial data and the complexity of spatial data types and spatial access methods. Challenges in spatial data mining arise from different issues. First, classical data mining is designed to process numbers and categories, whereas spatial data are more complex and include extended objects such as points, lines, and polygons. Second, while classical data mining works with explicit inputs, spatial predicates and attributes are often implicit. Third, classical data mining treats each input independently of other inputs, while spatial patterns often exhibit continuity and high autocorrelation among nearby features (Shekhar et al. 2002).

RELATED WORK

Statistical spatial data analysis has been a popular approach used to analyze spatial data. This approach handles numerical data well and usually suggests realistic models of spatial phenomena. Different methods for knowledge discovery, algorithms, and applications for spatial data mining are created. Classification of spatial data has been analyzed by some researchers. A method for classification of spatial objects has been proposed by Ester et al. (1997). Their proposed algorithm is based on ID3 algorithm, and it uses the concept of neighborhood graphs. It considers not only non-spatial properties of the classified objects, but also non-spatial properties of neighboring objects: objects are treated as neighbors if they satisfy the neighborhood relations. Ester et al. (2000) also define topological relations as those which are invariant under topological transformations. If both objects are rotated, translated, or scaled simultaneously, the relations are preserved. These scholars present a definition of topological relations derived from the nine intersections model: i.e. the topological relations between two objects are (1) disjoint, (2) meet, (3) overlap, (4) equal, (5) cover, (6) covered-by, (7) contain, and (8) inside; the second type of relations refers to (9) distance relations. These relations compare the distance between two objects with a given constant using arithmetic operators like <, >, and =. The distance between objects is defined as the minimum distance between them. The third type of relations they define are the direction relations. They define a direction relation *A R B* of two spatial objects using one representative point of the object A and all points of the destination object B. It is possible to define several possibilities of direction relations depending on the points that are considered in the source and the destination objects. The representative point of a source object may be the center of the object or a point on its boundary. The representative point is used as the origin of a virtual coordinate system, and its quadrants define the directions.

Fayyad et al. (1996) used decision tree methods to classify images of stellar objects to detect stars and galaxies. About three terabytes of sky images were analyzed. Similar to the mining association rules in transactional and relational databases, spatial association rules can be mined in spatial databases. Spatial association describes the spatial and non-spatial properties which are typical for the target objects but not for the whole database (Ester et al. 2000). Koperski et al. (1995) introduced spatial association rules that describe associations between objects based on spatial neighborhood relations. An example can be the following:

is_a(X,"African_countries")^receiving(X,"Western_aid")→
highly(X,"corrupt")[0.5%,90%]

This rule states that 90% of African countries receiving Western aid are also highly corrupt, and 0.5% of the data belongs to such a case.

Spatial clustering identifies clusters or densely populated regions according to some distance measurement in a large, multidimensional dataset. There are different methods for spatial clustering such as the k-mediod clustering algorithms (Ng et al. 1994) and the Generalized Density Based Spatial Clustering of Applications with Noise (GDBSCAN) that rely on a destiny-based notion of clusters (Sander et al. 1998).

Visualizing large spatial datasets became an important issue due to the rapidly growing volume of spatial datasets, which makes it difficult for a human to browse such datasets. Shekhar et al. (2002) have constructed a Web-based visualization software package for observing the summarization of spatial patterns and temporal trends. The visualization software will help users gain insight and enhance their understanding of the large data.

Mining Text Databases

Text databases consist of large collections of documents from various sources, such as news articles, research papers, books, digital libraries, E-mail messages, and Web pages. Text databases are rapidly growing due to the increasing amount of information available in electronic forms, such as electronic publications, E-mail, CD-ROMs, and the World Wide Web (which also can be considered as a huge interconnected dynamic text and multimedia database).

Data stored in most text databases are semi-structured data in that they are neither completely unstructured nor completely structured. For example, a document may contain a few structured fields, such as a title, author's name(s), publication date, length, category, etc., and also contain some largely unstructured text components, such as an abstract and contents.

Traditional information retrieval techniques have become inadequate for the increasingly vast amounts of text data (Han et al. 2001). Typically, only a small fraction of the many available documents will be relevant to a given individual user. Without knowing what could be in the documents, it is difficult to formulate effective queries for extracting and analyzing useful information from the data. Users need tools to compare different documents, rank the importance and relevance of the documents, or find patterns and trends across multiple documents. Thus, text mining has become an increasingly popular and essential theme in data mining.

Text mining has also emerged as a new research area of text processing. It focuses on the discovery of new facts and knowledge from large collections of texts that do not explicitly contain the knowledge to be discovered (Gomez et al. 2001). The goals of text mining are similar to those of data mining, since it attempts to find clusters, uncover trends, discover associations, and detect deviations in a large set of texts. Text mining has also adopted techniques and methods of data mining, such as statistical techniques and machine learning approaches. Text mining helps one to dig out the hidden gold from textual information, and it leaps from old fashioned information retrieval to information and knowledge discovery (Dorre et al. 1999).

Basic Measures for Text Retrieval

Information retrieval is a field that has been developing in parallel with database systems for many years. Unlike the field of database systems, however, which has focused on query and transaction processing of structured data, information retrieval is concerned with the organization and retrieval of information from a large number of text based documents. A typical information retrieval problem is to locate relevant documents based on user input, such as keywords or example documents. This type of information retrieval system includes online library catalog systems and online document management systems (Berry et al. 1999).

It is vital to know how accurate or correct a text retrieval system is in retrieving documents based on a query. The set of documents relevant to a query can be called "{Relevant}," whereas the set of documents retrieved is denoted as "{Retrieved}." The set of documents that are both relevant and retrieved is denoted as "{Relevant}∩{Retrieved}." There are two basic measures for assessing the quality of a retrieval system: (1) precision and (2) recall (Berry at al. 1999).

The precision of a system is the ratio of the number of relevant documents retrieved to the total number of documents retrieved. In other words, it is the percentage of retrieved documents that are in fact relevant to the query—i.e. the correct response. Precision can be represented as follows:

$$\text{Precision} = \frac{|\{Relevant\} \cap \{Retrieval\}|}{|\{Retrieved\}|}$$

The recall of a system is the ratio of the number of relevant documents retrieved to the total number of relevant documents in the collection. Stated differently, it is the percentage of documents that are relevant to the query and were retrieved. Recall can be represented the following way:

$$\text{Recall} = \frac{|\{Relevant\} \cap \{Retrieval\}|}{|\{Relevant\}|}$$

Word Similarity

Information retrieval systems support keyword-based and/or similarity-based retrieval. In keyword-based information retrieval, a document is represented by a string, which can be identified by a set of keywords. A good information retrieval system should consider synonyms when answering the queries. For example, synonyms such as "automobile" and "vehicle" should be considered when searching the keyword "car." There are two major difficulties with a keyword-based system: (1) synonymy and (2) polysemy. In a synonymy problem, keywords such as "software product" may not appear anywhere in a document, even though the document is closely related to a software product. In a polysemy problem, a keyword such as "regression" may mean different things in different contexts.

The similarity-based retrieval system finds similar documents based on a set of common keywords. The output of such retrieval should be based on the degree of relevance, where relevance is measured in terms of the closeness and relative frequency of the keywords.

A text retrieval system often associates a stop list with a set of documents. A stop list is a set of words that are deemed "irrelevant." For instance, "a," "the," "of," "for," "with," etc. are stop words, even though they may appear frequently. The stop list depends on the document itself: for example, together, "artificial intelligence" could be an important keyword in a newspaper; it may, however, be considered a stop word on research papers presented at an artificial intelligence conference.

A group of different words may share the same word stem. A text retrieval system needs to identify groups of words in which the words in a group are small syntactic variants of one another, and collect only the common word stem per group. For example, the group of words "drug," "drugged," and "drugs" share a common word stem, "drug," and can be viewed as different occurrences of the same word.

Panel et al. (2002) computed the similarity among a set of documents or between two words, w_i and w_j, using the cosine coefficient of their mutual information vectors:

$$\mathrm{sim}(w_i, w_j) = \frac{\sum_c {}^{mi}w_i c \times {}^{mi}w_j c}{\sqrt{\sum_c {}^{mi}w_i c^2 \times \sum_c {}^{mi}w_j c^2}}$$

where ${}^{mi}w_i c$ is the positive mutual information between context (c) and the word (w). $F_c(w)$ be the frequency count of the word, w, occurring in context c:

$$Mi_{wic} = \frac{\frac{F_c(w)}{N}}{\frac{\sum_i F_i(w)}{N} \times \frac{\sum_j F_c(J)}{N}}$$

Where

$$N = \sum_i \sum_j F_i(j)$$

is the total frequency counts of all words and their context.

Related Work

Text mining and applications of data mining to structured data derived from texts have been the subjects of many research projects in recent years. Most text mining has used natural language processing to extract key terms and phrases directly from documents.

Pantel at al. (2002) have proposed a clustering algorithm, Clustering By Committee (CBC), in which the centroid of a cluster is constructed by averaging the feature vectors of a subset of the cluster members. The subset is viewed as a committee that determines which other elements belong to the cluster. Pantel and his partners divided the algorithm into three phases. In the first phase, they found top similar elements. To compute the top similar words of a word *w*, they sorted *w*'s features according to their mutual information with *w*. They computed only the pairwise similarities between *w* and the words that share high mutual information features with *w*. In the second phase, they found committees—a set of recursively tight clusters in the similarity space—and other elements that are not covered by any committee. An element is said to be covered by a committee if that element's similarity to the centroid of the committee exceeds some high similarity threshold. Assigning elements to clusters is the last phase on the CBC algorithm. In this phase, every element is assigned to the cluster containing the committee to which it is most similar.

Wong et al. (1999) designed a text association system based on ideas from information retrieval and syntactic analysis. In their system, the corpus of narrative text is fed into a text engine for topic extractions, and then the mining engine reads the topics from the text engine and generates topic association rules which are sent to the visualization system for further analysis. There are two text engines developed on this system. The first one is word-based and results in a list of content-bearing words for the corpus. The second one is concept-based and results in concepts based on the corpus.

Dhillon et al. (2001) designed a vector space model to obtain a highly efficient process for clustering very large collections exceeding 100,000 documents in a reasonable amount of time on a single processor. They used efficient and scalable data structures such as local and global hash tables. In addition, a highly efficient and effective *spherical k-means* algorithm was used, since both the document and concept vectors lie on the surface of a high-dimensional sphere. The basic idea of the vector space model is to represent each document as a vector of certain weighted word frequencies. Each vector component reflects the importance of a particular term in representing the semantics or meaning of that document. The vectors for all documents in a database are stored as the columns of a single matrix (Berry et al. 1999).

A database containing a total of *d* documents described by *t* terms is represented as a *t x d term-by-document matrix A*. The *d* vectors representing the *d* documents form the columns of the matrix. Thus, the matrix element a_{ij} is the weighted frequency at which the term *i* occurs in document *j*. In the vector space model, the columns of *A* are the *document vectors*, whereas the rows of *A* are the *term vectors*.

To create the vector space model, there are important parsing and extraction steps needed, such as all unique words from the entire set of documents. Eliminate all "stop words" such as "a," "and," "the," etc. For each document, count the number of occurrences of each word. Eliminate "high-frequency" and "low-frequency" non-content-bearing words by using the heuristic or information-theoretic criteria. Finally, for each word, *w*, assign a unique identifier between one to *w* to each remaining word and unique identifier between one to *d* to each document. The geometric relationship between document vectors and also the term vectors can help to identify the similarities and differences between the document's content and also in term usage.

In the vector space model, in order to find the relevant documents, the user queries the database by using the vector space representation of those documents. Query matching is finding the documents most similar to the query in use and weighting the terms. In the vector space model, the documents selected are those geometrically closest to the query according to some measure.

Mining Remote Sensing Data

The data volumes of remote sensing are rapidly growing. National Aeronautics and Space Administration's (NASA) Earth Observing System (EOS) program alone produces massive data products with total rates more than 1.5 terabytes per day (King et al. 1999). Application and products of Earth observing and remote sensing technologies have been shown to be crucial to global social, economic and environmental well being (Yang et al. 2001).

In order to help scientists search massive remotely sensed databases and find data of interest to them, and then order the selected data sets or subsets, several information systems have been developed for data ordering purposes to face the challenges of the rapidly growing volumes of data, since the traditional method where a user downloads data and uses local tools to study the data residing on a local storage system is no longer helpful. To find interesting data, scientists need an effective and efficient way to search through the data. Metadata are provided in a database to support data searching by commonly used criteria such as spatial coverage, temporal coverage, spatial resolution, and temporal resolution. Since metadata search itself may still result in large amounts of data, some textual restrictions, such as keyword searches, could be employed for interdisciplinary researchers to select data of interest to them. The usual data selection procedure is to specify a spatial/temporal range and to see what datasets are available under those conditions.

Yang and his colleagues (1998) developed a distributed information system, the Seasonal International Earth Science Information Partner (SIESIP), which is a federated system that provides services for data searching, browsing, analyzing, and ordering. The system will provide not only data, but also data visualization, analysis, and user support capabilities.

The SIESIP system is a multi-tiered, client-server architecture, with three physical sites or nodes, distributing tasks in the areas of user service, access to data and information products, archiving as needed, ingest and interoperability options, and other aspects. This architecture can serve as a model for many distributed Earth system science data. There are three phases of user interaction with the data and information system; each phase can be followed by other phases or it can be conducted independently:

Phase 1, Metadata Access: using the metadata and browse images provided by the SIESIP system, the user browses the data holding. Metadata knowledge is incorporated into the system and a user can issue queries to explore this knowledge.

Phase 2, Data Discovery/Online Data Analysis: the user gets a quick estimate of the type and quality of data found in Phase 1. Analytical tools are then utilized as needed, such as statistical functions and visualization algorithms.

Phase 3, Data Order: after the user locates the dataset of interest, s/he is now ready to order datasets. If the data are available through SIESIP, the information system will handle the data order; otherwise, an order will be issued to the appropriate data provider such as Earth Science Distributed Active Archive Center (GES DAAC), on behalf of the user, or necessary information will be forwarded to the user for this task for further action.

The database management system is used for the system to handle catalogue metadata and statistical summary data. The database system supports two major kinds of queries. The first query is used to find the right data files for analysis and ordering based on catalogue metadata only. The second one queries data contents which are supported by the statistical summary data.

Data mining techniques help scientific data users not only in finding rules or relations among different data but also in finding the right datasets. With the rapid growth of massive data volumes, scientists need a fast way to search for the data in which they are interested. In this case, scientists need to search data based not only on metadata but also actual data values. The main goal of the data mining techniques in the SIESIP system is to find spatial regions and/or temporal ranges over which parameter values fall into certain ranges.

The main challenge is the speed and the accuracy, because they affect each other inversely.

Different data mining techniques can be applied on remote sensing data. Classification of remotely sensed data is used to assign corresponding levels with respect to groups with homogeneous characteristics, with the aim of discriminating multiple objects from one another within the image. Also, several methods exist for remote sensing image classification. Such methods include both traditional statistical or supervised approaches and unsupervised approaches that usually employ artificial neural networks.

MINING ASTRONOMICAL DATA

Astronomy has become and immensely data-rich field, with numerous digital sky surveys across a range of wavelengths and many terabytes of pixels and billions of detected sources, often with tens of measured parameters for each object. The problem with the astronomical database is not only the very large size, but also the variable quality of the data and the nature of astronomical objects with their very wide dynamic range in apparent luminosity and size present additional challenges. The great changes in astronomical data enable scientists to map the universe systematically, and in a panchromatic manner. Scientists can study the galaxy and the large scale structure in the universe statistically and discover unusual or new types of astronomical objects and phenomena (Brunner et al. 2002).

Astronomical data and their attendant analyses can be classified into the following five domains:

(1) *Imaging data* are the fundamental constituents of astronomical observations, capturing a two-dimensional spatial picture of the universe within a narrow wavelength region at a particular epoch or instance of time.
(2) *Catalogs* are generated by processing the imaging data. Each detected source can have a large number of measured parameters, including coordinates, various flux quantities, morphological information, and a real extant.
(3) *Spectroscopy, polarization, and other follow-up measurements* provide detailed physical quantification of the target systems, including distance information, chemical composition, and measurements of the physical fields present at the source.
(4) *Studying the time domain* provides important insights into the nature of the universe by identifying moving objects, variable sources, or transient objects.

(5) *Numerical simulations* are theoretical tools which can be compared with observational data.

Handling and exploring these vast new data volumes, and actually making real scientific discoveries, pose considerable technical challenges. Many of the necessary techniques and software packages, including artificial intelligence techniques like neural networks and decision trees, have been already successfully applied to astronomical problems such as pattern recognition and object classification. Clustering and data association algorithms have also been developed.

As early as 1936, Edwin Hubble established a system to classify galaxies into three fundamental types. First, elliptical galaxies had an elliptical shape with no other discernible structure. Second, spiral galaxies had an elliptical nucleus surrounded by a flattened disk of stars and dust containing a spiral pattern of brighter stars. And third, irregular galaxies have irregular shapes and did not fit into the other two categories.

Humphrey and his partners (2001) created an automated classification system for astronomical data. They visually classified 1,500 galaxy images obtained from the Automated Plate Scanner (APS) database in the region of the north galactic pole. Given the size and brightness of galaxy images taken into consideration, images that were difficult to classify were removed from this sample.

Grossman and company (2001) developed an application which simultaneously works with two geographically distributed astronomical source catalogs: (1) the Two Micron All Sky Survey (2MASS) and (2) the Digital Palomar Observatory Sky Survey (DPOSS). The 2MASS data are in the optical wavelengths, whereas the DPOCC data are in the infrared range. These scientists created a virtual observatory supporting the statistical analysis of many millions of stars and galaxies with data coming from both surveys. By using the data space transfer protocol (DSTP), a platform independent way to share data over a network, they built a query for finding all pairs from the DPOSS and 2MASS. The query visualized by coloring the data as red, if they appear in 2MASS; blue, if they appear in DPOSS; or magenta, if they appear in both surveys. The client DSTP application formulates the fuzzy join and sends the resulting stars and galaxies back to the client application. The DSTP protocol enables an application in one location to locate, access, and analyze data from several other locations.

Mining Bioinformatics Data

Bioinformatics is described by Cannataro and his colleagues (2004) as a bridge between the life sciences and computer science. It has also been described by

Barker and his associates (2004) as a cross-disciplinary field in which biologists, computer scientists, chemists, and mathematicians work together, each bringing a unique point of view (i.e. pluridisciplinary). The term *bioinformatics* has a range of interpretations, but the core activities of bioinformatics are widely acknowledged: storage, organization, retrieval and analysis of biological data obtained by experiments or by querying databases.

The increasing volume of biological data collected in recent years has prompted increasing demand for bioinformatics tools for genomic and proteomic (the set of proteins encoded by the genome to define models representing and analyzing the structure of the proteins contained in each cell) data analysis. Bioinformatics applications' design should represent the biological data and databases efficiently; contain services for data transformation and manipulation such as searching in protein databases, protein structure prediction, and biological data mining; describe the goals and requirements of applications and expected results; and support querying and computing optimization to deal with large datasets (Wong et al. 2001).

Bioinformatics applications are naturally distributed, due to the high number of datasets involved. They require higher computing power, due to the large size of datasets and the complexity of basic computations; they may access heterogeneous data; they require a secure software infrastructure because they could access private data owned by different organizations.

Cannataro et al. (2004) show technologies that can fulfill bioinformatics requirements. The following are some of these technologies:

(a) *Ontologies* are used to describe the semantics of data sources, software components and bioinformatics tasks. An *ontology* is a shared understanding of well defined domains of interest, which is realized as a set of classes or concepts, properties, functions and instances.
(b) *Workflow Management Systems* are employed to specify in an abstract way complex (distributed) applications, integrating and composing individual simple services. A workflow is a partial or total automation of a business process in which a collection of activities must be executed by some entities (humans or machines) according to certain procedural rules.
(c) *Grid Infrastructure* is used to show its security, distribution, service orientation, and computational power.
(d) *Problem Solving Environment* is used to define and execute complex applications, hiding software development details.

These researchers developed two types of ontologies: (1) *OnBrowser* for browsing and querying ontologies and (2) DAMON for the data mining domain describing resources and processes of knowledge discovery in databases. The latter is used to describe data mining experimentations on bioinformatics data.

Cannataro et al. (2004) also designed PROTEUS, a Grid-based Problem Solving Environment (GPSE), for composing and running bioinformatics applications on the Grid. They used ontologies for modeling bioinformatics processes and Grid resources and workflow techniques for designing and scheduling bioinformatics applications.

PROTEUS assists users in formulating bioinformatics solutions by choosing among different available bioinformatics applications or by composing a new one as collections on the Grid. It is used to present and analyze results and then compare them with past results to form the PROTEUS knowledge base. PROTEUS combines existing open source bioinformatics software and public-available biological databases by adding metadata to software, modeling applications through ontology and workflows, and offering pre-packaged Grid-aware bioinformatics applications.

Web-based bioinformatics application platforms are popular tools for biological data analysis within the bioscience community. Wong et al. (2001) developed a prototype based on integrating different biological databanks into a unified XML framework. The prototype simplifies the software development process of bioinformatics application platforms. The XML-based wrapper of the prototype demonstrated a way to convert data from different databanks into XML format and be stored in XML database management systems.

DNA data analysis is an important topic in biomedical research. Recent research in the area has led to the discovery of genetic causes for many diseases and disabilities, as well as the discovery of new medicines and approaches for disease diagnosis, prevention, and treatment. Data mining has become a powerful tool and contributes substantially to DNA analysis in the following ways, according to Han et al. (2001):

(a) *Semantic integration of heterogeneous, distributed genome database*: due to the highly distributed, uncontrolled generation and use of a wide variety of DNA data, the semantic integration of such heterogeneous and widely distributed genome databases becomes a pivotal task for systematic and coordinated analysis of DNA databases.
(b) *Similarity search and comparison among DNA sequences*: one of the most important search problems in genetic analysis is similarity search and comparison among DNA sequences. Gene sequences isolated from diseased and healthy species can be compared to identify critical differences between the two classes of genes. This can be done by first retrieving the gene sequences from the two tissue classes and then finding and comparing the frequently occurring patterns of each class.

(c) *Association analysis and identification of co-occurring gene sequences*: association analysis methods can be used to help determine the kinds of genes that are likely to co-occur in target samples. Such analysis would facilitate the discovery of groups of genes and the study of interactions and relationships between them.
(d) *Visualization tools and genetic data analysis*: complex structures and sequencing patterns of genes are most effectively presented in graphs, trees, cuboids, and chains by various kinds of visualization tools. Visualization, thus, plays an important role in biomedical data mining.

RESEARCH METHODOLOGY

The following is a discussion of the proposed approach for mining massive datasets for studying Africa from an Africancentric perspective. It depends on the METANET concept: a heterogeneous collection of scientific databases envisioned as a national and international digital data library which would be available via the Internet. I consider a heterogeneous collection of massive databases such as remote sensing data and text data. The discussion is divided into two separate, but interrelated, sections: (1) the automated generation of metadata and (2) the query and search of the metadata.

Automated Generation of Metadata

Metadata simply means data about data. They can be characterized as any information required to make other data useful in information systems. Metadata are a general notion that captures all kinds of information necessary to support the management, query, consistent use and understanding of data. Metadata help users to discover, understand and evaluate data, and help data administrators to manage data and control their access and use. Metadata describe how, when and by whom a particular set of data was collected, and how the data are formatted. Metadata are essential for understanding information stored in data warehouses.

In general, there exist metadata to describe file and variable types and organization, but they have minimal scientific content data. In raw form, a dataset and its metadata have minimal usability. For example, not all the image datasets in the same file form that are produced by the satellite-based remote sensing platform are important to scientists. In fact, only the image datasets that contain certain patterns will be of interest to a scientist. Scientists need metadata about the image datasets' content to enable scientists to narrow their search time, taking into consideration the size of the datasets: i.e. terabyte datasets (Wegman 1997).

Creating a digital object and linking it to the dataset will make the data usable, and at the same time the search operation for a particular structure in a dataset will be a simple indexing operation on the digital objects linked to the data. The objective of this process it to link digital objects with scientific meaning to the dataset at hand and make the digital objects part of the searchable metadata associated with the dataset. Digital objects will help scientists to narrow the scope of the datasets that the scientists must consider. In fact, digital objects reflect the scientific content of the data, but they do not replace the judgment of the scientists.

The digital objects will essentially be named for patterns to be found in the datasets. The goal is to have a background process, launched either by the database owner or, more likely, via an applet created by a virtual data center, examining databases available on the data-Web and searching within datasets for recognizable patterns. Once a pattern is found in a particular dataset, the digital object corresponding to that pattern is made part of the metadata associated with that set. If the same pattern is contained in other distributed databases, pointers would be added to that metadata pointing to metadata associated with the distributed databases. The distributed databases will then be linked through the metadata in the virtual data center.

At least one of the following three different methods is to be used to generate the patterns for which a researcher can search. The first method is to delineate empirical or statistical patterns that have been observed over a long period of time and may be thought to have some underlying statistical structure. An example of an empirical or statistical pattern is a certain pattern in DNA sequencing. The second method is to generate the model-based patterns. This method is predictive if verified on real data. The third method is to tease out patterns found by clustering algorithms. With this method, patterns are delineated by purely automated techniques that may or may not have scientific significance (Wegman 1997).

Query and Search

The notion of the automated creation of metadata is to develop metadata that reflect the scientific content of the datasets within the database rather than just data structure information. The locus of the metadata is the virtual data center where it is reproduced. The general *desideratum* for the scientist is to have a comparatively vague question that can be sharpened as s/he interacts with the system. Scientists can create a query, and this query may be sharpened to another vague one, but the data may be accessible from several distributed databases for the later query. The main issues of the retrieval process are the browser mechanism for requesting data when the user has a precise query and an expert system query capability that would help the scientist reformulate a vague question into a form that may be submitted more precisely.

Query and search would comprise four major elements: (1) a client browser, (2) an expert system for query refinement, (3) a search engine, and (4) a reporting mechanism. These four elements are described in the following subsections.

Client Browser

The client browser would be a piece of software running on a scientist's client machine, which is likely to be a personal computer (PC) or a workstation. The main idea is to have a graphical user interface (GUI) that would allow the user to interact with a more powerful server in the virtual data center. The client software is essentially analogous to the myriad of browsers available for the World Wide Web (WWW).

Expert Systems for Query Refinement

A scientist interacts with a server via two different scenarios. In the first scenario, the scientist knows precisely the location and type of data s/he desires. In the second scenario, the scientist knows generally the type of questions s/he would like to ask, but has little information about the nature of the databases with which s/he hopes to interact. The first scenario is relatively straightforward, but the expert system would sill be employed to keep a record of the nature of the query. The idea is to use the query as a tool in the refinement of the search process. The second scenario is more complex. The approach is to match a vague query formulated by the scientist to one or more of the digital objects discovered in the automated generation of metadata phase. Disciplined experts give rules to the expert system to perform this match. The expert system would attempt to match the query to one or more digital objects. The scientist has the opportunity to confirm the match when s/he is satisfied with the proposed match or to refine the query. The expert system would then engage the search engine in order to synthesize the appropriate datasets. The expert system would also take advantage of the interaction to form a new rule for matching the original query to the digital objects developed in the refinement process. Thus, two aspects emerge: (1) the refinement of the precision of an individual search and (2) the refinement of the search process. Both aspects share tactical and strategic goals. The refinement would be greatly aided by the active involvement of the scientist. S/he would be informed about how his/her particular query was resolved, allowing him/her to reformulate the query efficiently. The log files of these iterative queries would be processed automatically to inspect the query trees and, possibly, improve their structure.

Also, two other considerations of interest emerge. First, other experts not necessarily associated with the data repository itself may have examined certain datasets and have commentaries in either informal annotations or in the refereed scientific literature. These commentaries should form part of the metadata associated with the dataset. Part of the expert system should provide an annotation mechanism that would allow users to attach commentaries or library references (particularly digital library references) as metadata. Obviously, such annotations may be self-serving and potentially unreliable. Nonetheless, the idea is to alert the scientist to information that may be useful. User derived metadata would be considered secondary metadata.

The second consideration is to provide a mechanism for indicating data reliability. This would be attached to a dataset as metadata, but it may in fact be derived from the original metadata. For example, a particular data collection instrument may be known to have a high variability. Thus, any set of data that is collected by this instrument, no matter where in the data it occurred, should have as part of the attached metadata an appropriate caveat. Hence, an automated metadata collection technique should be capable of not only examining the basic data for patterns, but also examining the metadata themselves; and, based on collateral information such as just mentioned, it should be able to generate additional metadata.

Search Engine

As noted earlier, large scale scientific information systems will likely be distributed in nature and contain not only the basic data, but also structured metadata: for example, sensor type, sensor number, measurement data, and unstructured metadata such as a text-based description of the data. These systems will typically have multiple main repository sites that together will house a major portion of the data as well as some smaller sites and virtual data centers containing the remainder of the data. Clearly, given the volume of the data, particularly within the main servers, high performance engines that integrate the processing of the structured and unstructured data are necessary to support desired response rates for user requests.

Both database management systems (DBMS) and information retrieval systems provide some functionality to maintain data. DBMS allow users to store unstructured data as binary large objects (BLOB), and information retrieval systems allow users to enter structured data in zoned fields. DBMS, however, offer only a limited query language for values that occur in BLOB attributes. Similarly, information retrieval systems lack robust functionality for zoned fields. Additionally, information retrieval systems traditionally lack efficient parallel algorithms. Using a relational database approach for information retrieval allows for parallel processing, since almost all commercially available parallel engines support some relational DBMS. An inverted index may be modeled as a relation. This treats information retrieval as an application of a DBMS. Using this approach, it is possible to implement a variety of information retrieval functionality and achieve good run-time performance. Users can issue complex queries including both structured data and text.

The key hypothesis is that the use of a relational DBMS to model an inverted index will (a) permit users to query both structured data and text via standard Structured Query Language (SQL)—in this regard, users may use any relational DBMS that support standard SQL; (b) permit the implementation of traditional information retrieval functionality such as Boolean retrieval, proximity searches, and relevance ranking, as well as non-traditional approaches based on data fusion and machine learning techniques; and (c) take advantage of current parallel DBMS implementations, so that acceptable run-time performance can be obtained by increasing the number of processors applied to the problem.

Reporting Mechanism

The most important issue for a reporting mechanism is not only to retrieve datasets appropriate to the needs of the scientist, but scaling down the potentially large databases the scientist must consider. Put differently, the scientist would consider megabytes ($\sim 10^6$ bytes) instead of terabytes ($\sim 10^{12}$ bytes) of data. The search and retrieval process may still result in a massive amount of data. The reporting mechanism would, therefore, initially report the nature and magnitude of the datasets to be retrieved. If the scientist agrees that the scale is appropriate for his/her needs, then the data will be delivered by a file transfer protocol (FTP) or a similar mechanism to his/her local client machine or to another server where s/he wants the synthesized data to be stored.

IMPLEMENTATION

To help scientists search for massive databases and find data of interest to them, a good information system should be developed for data ordering purposes. The system should be performing well based on the descriptive information of the scientific datasets or metadata, such as the main purpose of the datasets, the spatial and temporal coverage, the production time, the quality of the datasets, and the main features of the datasets.

Scientists want to have an idea of what the data look like before ordering them, since metadata searching alone cannot meet all scientific queries. Thus, content-based searching or browsing and preliminary analysis of data based on their actual values will be inevitable in these application contexts. One of the most common content-based queries is to find large enough spatial regions over which the geophysical parameter values fall into certain intervals given a specific observation time. The query result could be used for ordering data as well as for defining features associated with scientific concepts.

For researchers of African topics to be able to maximize the utility of this content-based query technique, there must exist a Web-based prototype through which they can demonstrate the idea of interest. The prototype must deal with different types of massive databases, with special attention being given to the following and other aspects that are unique to Africa:

(a) African languages with words encompassing diacritical marks (dead and alive)
(b) Western colonial languages (dead and alive)
(c) Other languages such as Arabic, Russian, Hebrew, Chinese, etc.
(d) Use of desktop software such as Microsoft Word or Corel WordPerfect to type words with diacritical marks and then copy and paste them into Internet search lines
(e) Copying text in online translation sites and translating them into the target language

The underlying approach must be pluridiscipinary, which involves the use of open and resource-based techniques available in the actual situation. It has, therefore, to draw upon the indigenous knowledge materials available in the locality and make maximum use of them. Indigenous languages are, therefore, at the center of the effective use of this methodology (see Chapter 10 for more on this).

In the collection of remote sensing and text databases, one must implement a prototype system that contains at least a four-terabyte storage capability with high performance computing. Remote sensing data are available through NASA JPL, NASA Goddard, and NASA Langley Research Center.

The prototype system will allow scientists to make queries against disparate types of databases. For instance, queries on remote sensing data can focus on the features observed in images. Those features may be environmental or artificial features which consist of points, lines, or areas. Recognizing features is the key to interpretation and information extraction. Images differ in their features, such as tone, shape, size, pattern, texture, shadow, association, etc.

Tone refers to the relative brightness or color objects in the image. It is the fundamental element for distinguishing between different targets or features. Shape refers to the general form, structure, or outline of an object. Shape can be a very distinctive clue for interpretation. Size of objects in an image is a function of scale. It is important to assess the size of a target relative to other objects in a scene, as well as the absolute size, to aid in the interpretation of that target. Pattern refers to the spatial arrangement of visibly discernible objects. Texture refers to the arrangement and frequency of tonal variation in a particular area of an image. Shadow will help in the interpretation by providing an idea of the profile and relative height of a target or targets which may make identification easier. Association takes into account the relationship among other recognizable objects or features in proximity to the target of interest.

Other features of the images that also should be taken into consideration include percentage of water, green land, cloud forms, snow, and so on. The prototype system will help scientists to retrieve images that contain different features; the system should be able to handle complex queries. This calls for some knowledge of African fractals, which Ron Eglash (1999) and I (Bangura 2007) have defined as a self-similar pattern—i.e. a pattern that repeats itself on an ever diminishing scale (see also Chapter 12 for details).

As Ron Eglash (1999) has demonstrated, first, traditional African settlements typically show repetition of similar patterns at ever-diminishing scales: circles of circles of circular dwellings, rectangular walls enclosing ever-smaller rectangles, and streets in which broad avenues branch down to tiny footpaths with striking geometric repetition. He easily identified the fractal structure when he compared aerial views of African villages and cities with corresponding fractal graphics simulations. To estimate the fractal dimension of a spatial pattern, Eglash used several different approaches. In the case of Mokoulek, for instance, which is a black-and-white architectural diagram, a two-dimensional version of the ruler size versus length plots were employed. For the aerial photo of Labbazanga, however, an image in shades of gray, a Fourier transform was used (refer to Chapter 12 for the Fourier transform definition and equation). Nonetheless, according to Eglash, we cannot just assume that African fractals show an understanding of fractal geometry, nor can we dismiss that possibility. Thus, he insisted that we listen to what the designers and users of these structures have to say about it. This is because what may appear to be an unconscious or accidental pattern might actually have an intentional mathematical component.

Second, as Eglash examined African designs and knowledge systems, five essential components (recursion, scaling, self-similarity, infinity, and fractional dimension) kept him on track of what does or does not match fractal geometry. Since scaling and self-similarity are descriptive characteristics, his first step was to look for the properties in African designs. Once he established that theme, he then asked whether or not these concepts had been intentionally applied, and started to look for the other three essential components. He found the clearest illustrations of indigenous self-similar designs in African architecture.

The examples of scaling designs Eglash provided vary greatly in purpose, pattern, and method. As he explained, while it is not difficult to invent explanations based on unconscious social forces—for example, the flexibility in conforming designs to material surfaces as expressions of social flexibility—he did not believe that any such explanation can account for its diversity. He found that from optimization engineering, to modeling organic life, to mapping between different spatial structures, African artisans had developed a wide range of tools, techniques, and design practices based on the conscious application of scaling geometry. Thus, for example, instead of using the Koch curve to generate the branching fractals used to model the lungs and acacia tree, Eglash used passive lines that are just carried through the iterations without change, in addition to active lines that create a growing tip by the usual recursive replacement.

For the text database, the prototype system must consider polysymy and synonymy problems in the queries. Polysymy means words having multiple meanings: e.g. "order," "loyalty," and "ally." Synonymy means multiple words having the same meaning: e.g., "jungle" and "forest," "tribe" and "ethnic-group," "language" and "dialect," "tradition" and "primitive," "corruption" and "lobbying."

The collected documents will be placed into categories depending on the documents' subjects. Scientists can search into those documents and retrieve only the ones related to queries of interest. Scientists can search via words or terms, and then retrieve documents on the same category or from different categories as long as they are related to the words or terms in which the scientists are interested.

CONCLUSION

Data mining techniques and visualization must play a pivotal role in retrieving substantive electronic data to study and teach about African phenomena in order to discover unexpected correlations and causal relationships, and understand structures and patterns in massive data. Data mining is a process for extracting implicit, nontrivial, previously unknown and potentially useful information such as knowledge rules, constraints, and regularities from data in massive databases. The goals of data mining are (a) explanatory—to analyze some observed events, (b) confirmatory—to confirm a hypothesis, and (c) exploratory—to analyze data for new or unexpected relationships. Typical tasks for which data mining techniques are often used include clustering, classification, generalization, and prediction. The most popular methods include decision trees, value prediction, and association rules often used for classification. Artificial Neural Networks are particularly useful for exploratory analysis as non-linear clustering and classification techniques. The algorithms used in data mining are often integrated into Knowledge Discovery in Databases (KDD)—a larger framework that aims at finding new knowledge from large databases. While data mining deals with transforming data into information or facts, KDD is a higher-level process using information derived from a data mining process to turn it into knowledge or integrate it into prior knowledge. In general, KDD stands for discovering and visualizing the regularities, structures and rules from data; discovering useful knowledge from data; and for finding new knowledge.

Visualization is a key process in Visual Data Mining (VDM). Visualization techniques can provide a clearer and more detailed view on different aspects of the data as well as results of automated mining algorithms. The exploration of relationships between several information objects, which represent a selection of the information content, is an important task in VDM. Such relations can either be given explicitly, when being specified in the data, or they can be given implicitly, when the relationships are the result of an automated mining process: for example, when relationships are based on the similarity of information objects derived by hierarchical clustering.

Understanding and trust are two major aspects of data visualization. Understanding is undoubtedly the most fundamental motivation behind visualizing massive data. If scientists understand what has been discovered from data, then they can trust the data. To help scientists understand and trust the implicit data discovered and useful knowledge from massive datasets concerning Africa, it is imperative to present the data in various forms, such as boxplots, scatter plots, 3-D cubes, data distribution charts, as well as decision trees, association rules, clusters, outliers, generalized rules, etc.

The software called Crystal Vision is a good tool for visualizing data. It is an easy to use, self-contained Windows application designed as a platform for multivariate data visualization and exploration. It is intended to be robust and intuitive. Its features include scatter plot matrix views, parallel coordinate views, rotating 3-D scatter plot views, density plots, multidimensional grand tours implemented in all views, stereoscopic capability, saturation brushing, and data editing tools. It has been used successfully with datasets as high as 20 dimensions and with as many as 500,000 observations (Wegman 2003). Crystal Vision is available at the following Internet site: <ftp://www.galaxy.gmu.edu/pub/software/CrystalVisionDemo.exe>

REFERENCES

Alshameri, Faleh J. 2006. Automated generation of metadata for mining image and text data. Doctoral dissertation, George Mason University, Fairfax, Virginia.

Bangura, Abdul Karim. 2005. Ubuntugogy: An African educational paradigm that transcends pedagogy, andragogy, ergonagy and heutagogy. *Journal of Third World Studies* xxii, 2:13–54.

Bangura, Abdul Karim. 2000. *Chaos Theory and African Fractals*. Washington, DC: The African Institution Publications.

Bangura, Abdul Karim. 2000. Book Review of Ron Eglash's *African Fractals: Modern Computing and Indigenous Design. Nexus Network Journal* 2, 4.

Barker, J. and J. Thornton. 2004. Software engineering challenges in bioinformatics. *Proceedings of the 26th International Conference on Software Engineering* (ICSE '04).

Berry, M., Z. Drmac and E. Jessup. 1999. Matrices, vector spaces, and information retrieval. *Society for Industrial Applied Mathematics* (SIAM) 41, 2:335–362.

Brunner, R., S. Djorgovsky, T. Prince and A. Szalay. 2002. Massive datasets in astronomy. In J. Abello et al., eds. *Handbook of Massive Datasets*. Norwell, MA: Kluwer Academic Publishers.

Cannataro, M., C. Comito, A. Guzzo and P. Veltri. 2004. Integrating ontology and workflow in PROTEUA, a grid-based problem solving environment for bioinformatics. *Proceedings of the International Conference on Information Technology: Coding and Computing* (ITCC '04).

Chen, M., J. Han and P. Yu. 1996. Data mining: An overview from a database perspective. *IEEE Transactions on Knowledge and Data Engineering* 8, 6:866–883.

Dhillon, I., J. Han and Y. Guan. 2001. Efficient clustering of very large document collection. In R. Grossman et al., eds. *Data Mining for Scientific and Engineering Applications*. Norwell, MA: Kluwer Academic Publishers.

Dorre, J., P. Gerstl and R. Seiffert. 1999. Text mining: Finding nuggets in mountains of textual data. KDD-99:398–401. San Diego, CA.

Eglash, Ron. 1999. *African Fractals: Modern Computing and Indigenous Design*. New Brunswick, NJ: Rutgers University Press.

Ester, M., H. Kriegel and J. Sander. 2001. Algorithms and applications for spatial data mining. *Geographic Data Mining and Knowledge Discovery, Research Monographs in GIS*. Taylor and Francis, 167–187.

Ester, M., H. Kriegel and J. Sander. 1997. Spatial data mining: A database approach. *Proceedings of the International Symposium on Large Spatial Databases*. SSD '97:47–66. Berlin, Germany.

Ester, M., A. Frommelt, H. Kriegel and J. Sander. 2000. Spatial data mining: Database primitives, algorithms and efficient DBMS support. *Data Mining and Knowledge Discovery* 4, 2/3:193–216.

Fayyad U. M., G. Piatetsky-Shapiro and P. Smyth. 1996. From data mining to knowledge discovery: An overview. U. M. Fayyad et al., eds. *Advances in Knowledge Discovery and Data Mining*. Menlo Park, CA: AAAI Press.

Gomez, M., A. Gelbuhk, A. Lopez and R. Yates. 2001. Text mining with conceptual graphs. *IEEE* 893–903.

Grossman, R., E. Creel, M. Mazzucco and R. Williams. 2001. A dataspace infrastructure for astronomical data. In R. Grossman et al., eds. *Data Mining for Scientific and Engineering Applications*. Norwell, MA: Kluwer Academic Publishers.

Hambrusch, S., C. Hoffman, M. Bock, S. King and D. Miller. 2003. Massive data: Management, analysis, visualization, and security. A School of Science Focus Area at Purdue University Report.

Han, J. and M. Kamber. 2001. *Data Mining: Concepts and Techniques*. San Francisco, CA: Morgan Kaufman Publishers.

Humphreys, R., J. Cabanela and J. Kriessler. 2001. Mining astronomical databases. In R. Grossman et al., eds. *Data Mining for Scientific and Engineering Applications*. Norwell, MA: Kluwer Academic Publishers.

Kafatos, M., R. Yang, X. Wang, Z. Li and D. Ziskin. 1998. Information technology implementation for a distributed data system serving earth scientists: Seasonal to International ESIP. *Proceedings of the 10th International Conference on Scientific and Statistical Database Management* 210–215.

Kamath, C. 2001. On mining scientific datasets. In R. Grossman et al., eds. *Data Mining for Scientific and Engineering Applications*. Norwell, MA: Kluwer Academic Publishers.

King, M. and R, Greenstone. 1999. *EOS Reference Handbook*. Washington, DC: NASA Publications.

Koperski, K. and J. Han. 1995. Discovery of spatial association rules in geographic information databases. *Proceedings of the 4th International Symposium on Advances in Spatial Databases*. 47–66. Portland, ME.

Koperski, K. and J. Han and N. Stefanovic. 1998. An efficient two-step method for classification of spatial data. *Proceedings of the Symposium on Spatial Data Handling*. 45–54. Vancouver, Canada.

Marusic, I., G. Candler, V. Interrante, P. Subbareddy and A. Moss. 2001. Real time feature extraction for the analysis of turbulent flows. In R. Grossman et al., eds. Data Mining for Scientific and Engineering Applications. Norwell, MA: Kluwer Academic Publishers.

Ng, R. and J. Han. 1994. Efficient and effective clustering methods for spatial data mining. *Proceedings of the 20th International Conference on Very Large Databases*. 144–155, Santiago, Chile.

Palacio, M., D. Sol and J. Gonzalez, 2003. Graph-based knowledge representation for GIS data. *Proceedings of the Fourth Mexican International Conference on Computer Science* (ENC '03).

Pantel, P. and D. Lin. 2002. Discovering word senses from text. In *Proceedings of SIGKDD-01*. San Frncisco, CA.

Sander, J. M. Ester and H. Kriegel. 1998. Density-based clustering in spatial databases: A new algorithm and its applications. *Data Mining and Knowledge Discovery* 2, 2:169–194.

Shekhar, S., C. Lu, P. Zhang, and R. Liu. 2002. Data mining and selective visualization of large spatial datasets. *Proceedings of the 14th IEEE International Conference on Tools with Artificial Intelligence* (ICTAI '02).

Wegman, E. 2003. Visual data mining. *Statistics in Medicine* 22:1383-1397 plus 10 color plates.

Wegman, E. 1997. A Guide to Statistical Software. Available at http://www.galaxy.gmu.edu/papers/astr1.html

Wong, P., P. Whitney and J. Thomas. 1999. Visualizing association rules for text mining. In G. Wills and D, Keim, eds. *Proceedings of IEEE Information Visualization '99*. Los Alamitos, CA: IEEE CS Press.

Wong, R. and W. Shui. 2001. Utilizing multiple bioinformatics information sources: An XML database approach. *Proceedings of the Second IEEE International Symposium on Bioinformatics and Bioengineering*.

Yang, R., X. Deng, M. Kafatos, C. Wang and X. Wang. 2001. An XML-based distributed metadata server (DIMES) supporting earth science metadata. *Proceedings of the 13th International Conference on Scientific and Statistical Database Management* 251–256.

ACKNOWLEDGMENT

This chapter benefited greatly from the work of my colleague, Professor Faleh J. Alshameri.

Chapter Fourteen

General Conclusion: Access to Mathematics versus Access to the Language of Power: Lessons from the Struggle in South African Multilingual Mathematics Classrooms

INTRODUCTION

> Classroom conversations that include the use of . . . the [bilingual] students' first language as legitimate resources can support students in learning to communicate mathematically (Moschkovich, 2002: 208).

> If we changed our (Mathematics) textbooks into Setswana and set our exams in Setswana, then my school will be empty because our parents now believe in English (Lindi, a Grade 4 Mathematics teacher).

This concluding chapter explores two critical aspects of the development of Mathematics in Africa. The first aspect deals with how teachers and learners position themselves in relation to the use of language(s) in multilingual Mathematics classrooms. The examples we draw are from two studies in multilingual Mathematics classrooms in South Africa. The analysis presented shows that teachers and learners who position themselves in relation to English are concerned with access to social goods and positioned by the social and economic power of English (as is the case throughout Africa where other colonial languages still dominate in official matters). They do not focus on epistemological access but argue for English as the language of learning and teaching. In contrast, learners who position themselves in relation to Mathematics and, thus, epistemological access, reflect more contradictory discourses, including support for the use of the their home languages as languages of learning and teaching. The second aspect is the challenge that the Ethnomathematics strand poses for Mathematics education. In dealing with this issue, we focus on the work of one of its staunch proponents, Arthur Powell.

It is widely accepted that language is important for learning and thinking and that the ability to communicate mathematically is central to learning and teaching school Mathematics. What is still under constant debate and investigation is which language is most appropriate for learning a subject such as Mathematics, especially in a multilingual contexts. The preceding quotes capture the essence and complexity of the debate. Some researchers argue that the learners' main languages are a resource in the teaching and learning of Mathematics while some teachers argue for the use of English. Herein lies the heart of the problem explored in this chapter. These arguments are equally compelling, as they are about access to Mathematics and social goods. A major aim of this chapter is to give substance to the debate by exploring how multilingual Mathematics teachers and learners position themselves in this debate and what this might mean for research and practice.

The data used in this discussion are drawn from two research projects in multilingual Mathematics classrooms in South Africa. Using data from South Africa is convenient but also appropriate: South Africa is an extraordinarily complex multilingual country. While the multilingual nature of South African Mathematics classrooms may seem exaggerated, they are not atypical. In South Africa, there is a general view that most parents want their children to be educated in English and that most learners would like to be taught in English. While there is no systematic research evidence, it is also widely held that many schools with an African student body choose to use English as a language of learning and teaching (LoLT) from the first year of schooling (Taylor & Vinjevold, 1999). The TIMSS results in South Africa were very poor. Studies that have emerged from this argue that the solution to improving African learners' performance in Mathematics is to develop their English language proficiency (e.g., Howie, 2002). What does this recommendation mean for Mathematics learning? The question explored here is about how the power dynamics of language play out in the Mathematics classroom context, and in whose or what interests. Issues of power and access are by no means straightforward, and so it is important that they be problematized.

The work on the politics of language is complex, not well developed in Mathematics education, and often misrepresented. To put this debate in perspective it is important to provide a brief overview on the political role of language.

THE POLITICAL ROLE OF LANGUAGE AND ITS USE IN MULTILINGUAL MATHEMATICS CONTEXTS

Language, like multilingualism, is always political (Hartshone, 1987; Reagan & Ntshoe 1987; Mda, 1997; Friedman, 1997; Heugh, 1997; Granville et al.,

1998; Gee, 1999). It is one of the characteristics that are used in society to determine power (Gutiérrez, 2002). In South Africa, the issue of language has always been interwoven with the politics of domination and separation, resistance and affirmation. During apartheid, the language of learning issue became a dominating factor in opposition to the system of Bantu Education. Although not unmindful or ashamed of African traditions per se, the mainstream African nationalists have generally viewed cultural assimilation as a means by which Africans could be released from a subordinate position in a common, unified society (Reagan and Ntshoe, 1987). They therefore fought against the use of African languages as languages of learning and teaching because they saw them as a device to ensure that Africans remain oppressed. Lindi's views in the preceding quote that the parents of learners in her school believe in English are, thus, not surprising.

The political nature of language is not only at the macro-level of structures but also at the micro-level of classroom interactions. Language can be used to exclude or include people in conversations and decision-making processes. Zentella (1997) through her work with Puerto Rican children in El Bario, New York shows how language can bring people together or separate them. Language is one way in which one can define one's adherence to group values. Therefore, decisions about which language to use in multilingual Mathematics classrooms, how, and for what purposes, are not only pedagogic but also political (Setati, 2005). Most research on Mathematics education in multilingual classrooms has argued for the use of the learners' home languages as resources for learning and teaching Mathematics (e.g., Addendorff, 1993; Adler, 2001; Arthur, 1994; Khisty, 1995; Merrit, Cleghorn, Abagi and Bunyi, 1992; Moschkovich, 1999 & 2002; Setati and Adler, 2001). They have argued for the use of the learners' home languages in learning and teaching Mathematics as a support needed while learners continue to develop proficiency in the language of learning and teaching (e.g., English) at the same time as learning Mathematics. While research in general education on language and minority learners is strongly rooted in the socio-political context of learning (Cummins, 2000), most research on multilingualism in Mathematics education has been framed by a limited conception of language as a tool for thinking and communication. To ignore the political role of language in Mathematics education research and practice would assume that power relationships do not exist in society.

In this vein, we use the work of Gee (1996 & 1999) to take the work on multilingualism in Mathematics education further by explaining the language choices of teachers and learners in multilingual Mathematics classrooms beyond the pedagogic and cognitive. Gee's work is relevant because he considers language as always political (1996 & 1999). He argues that when people speak or write, they create a political perspective; they use language to project themselves as certain kinds of people engaged in certain kinds of activities. Language is therefore never just a vehicle to express ideas (a cultural tool), it is also a political tool that people use to enact (i.e. to be recognized as) a particular 'who' (identity) engaged in a particular 'what' (situated activity).

Gee uses the theoretical construct of cultural models to explore the identities and activities that people are enacting. Cultural models are shared, conventional ideas about how the world works, which individuals learn by talking and acting with their fellows. They help us explain why people do things in the way that they do and provide a framework for organizing and reconstructing memories of experience (Holland and Quinn, 1987). Cultural models do not reside in people's heads; they are embedded in words, in people's practices, and in the context in which they live. The issue that is relevant here is about the cultural models teachers and learners in multilingual Mathematics classrooms enact in relation to language and Mathematics. In what follows, we use the notion of cultural models to explore why teachers and learners prefer the language(s) that they choose for learning and teaching Mathematics. Thereafter, we look at the implications of such language choices for research and practice.

TEACHERS' LANGUAGE CHOICES

The data that we draw upon in this section come from observations that involved six primary school Mathematics teachers in multilingual classrooms in South Africa. The data were collected through individual teacher interviews, focus group interviews, and classroom observations. During the pre-observation interviews, teachers were asked: "In which language do you prefer to teach Mathematics? Why?" Over and above all else, *English is international* emerged as a dominant cultural model that shaped the teachers' language choices. All six teachers stated ideological and pragmatic reasons for their preference to teach Mathematics in English. As the following extracts show, these reasons ranged from the belief that English is an international language to the fact that textbooks, examinations and higher education are all in English.

> Vusi: I prefer to teach in English because it is a *universal language*.
>
> Kuki: I think all the languages must be equal; although English is the *international language*, it has to still be emphasised and mother tongue I think it's high time that the kids *learn mother tongue and be proud of it.*

> Lindi: . . . it is said that [English] is an *international language* . . . I encourage them to use English . . . The *textbooks are written in English,* the *question papers are in English*, so you find that the child doesn't understand what is written there (my emphasis).

These teachers are aware of the linguistic capital of English and the symbolic power it bestows on those who can communicate in it. They see English as international and universal and, thus, 'bigger than' themselves. The way Kuki and Lindi express themselves in the above extracts also suggests that they do not want to take responsibility for the status of English. The status of English is what it is and they cannot change it. Kuki uses the phrase "I think . . . ," while Lindi uses "It is said . . . ," suggesting that they see themselves as being caught up in the dominance of English. They do not have any control over the international nature of English. All they can do is to prepare their learners for participation in the international world, and teaching Mathematics in English is an important part of this preparation. It is thus not surprising that all the teachers saw English as the natural choice for use in Mathematics teaching. Throughout their lives, they have lived in an environment that values English more than any other language. Furthermore, as Lindi points out, the Mathematics textbooks and examinations are in English. Over the years, no Mathematics textbook in South Africa was written in an African language. During the time when 'mother tongue' instruction was enforced in primary schools, the Mathematics textbooks at this level were translated from English or Afrikaans into the African languages. The secondary school Mathematics textbooks have never been published in African languages in South Africa. Therefore, for many African teachers and learners, Mathematics is associated with the English language because it is the language of Mathematics textbooks. As a result, English has become the natural choice for teaching and learning Mathematics. What is interesting is that none of the teachers challenged the power of English or the fact that textbooks and examinations are in English while learners are not fluent in it.

While the other three teachers did not explicitly highlight the international nature of English, they also indicated that they encourage their learners to use English and their reasons focused on the social goods that learners can access through English.

> Gugu: I think English, *it empowers* them [the learners], you understand. At this stage of eight, nine years, they can be able to *speak English unlike us. We never did English in primary* and at college we were supposed to answer in English in lectures. So we had a problem with this language, *so at any early age they just become used to it.*

> Mpule: I *encourage them to use English* because if they do not learn the language how will they be able to *cope in higher classes*, they will not cope.
>
> Rosina: I encourage them to use *English always* . . . So that they can *learn the language* (my emphasis).

Gugu wants to make English accessible to her learners early in their schooling. In her view, making English accessible will assist in undoing the wrongs of the past, which she experienced as a learner. Gugu's view of making English accessible is similar to Granville et al.'s (1998), who insist that all South African learners must learn at least one African language. They argue that all school learners should have access to English, which is "the language of power" at present. The argument is that if everyone had access to English; it would no longer be an elitist language. In this way English could come to be seen as a resource, not as a "problem" (Granville et. al., 1998). The challenge now is that even the learners who do not have access to English are learning Mathematics in English. Gugu's view is that the Mathematics classroom is another opportunity for learners to gain access to English. An important question to ask here is this: What is the cost of focusing on making English accessible to learners during Mathematics teaching?

Mpule highlights the fact that English is the language of higher education. Higher education in South Africa is only available in English and Afrikaans. As a primary school teacher, she feels responsible for ensuring that her students are ready for higher classes and the ability to speak English is an important part of preparation for that. What is interesting is that Mpule, like all the other teachers in the study, does not highlight the importance of ensuring that learners are mathematically competent for higher classes. While this absence of a concern for mathematical competence may not be deliberate, it is important to note. What is in the foreground in the teachers' cultural models is English. Explanations for their preferred language(s) for Mathematics teaching focus on English and not Mathematics. These teachers position themselves in relation to English (and so socio-economic access) and not Mathematics (i.e. epistemological access).

Of all the teachers, Kuki is the only one who indicated some awareness of the fact that all the official languages in South Africa are equal. What is interesting is that even with this recognition, Kuki still maintains that English has to be emphasised. As the preceding extracts show, Kuki is working with conflicting cultural models of wanting to honour the African languages, on the one hand, and, on the other hand, ensuring that the learners have access to English. During the focus group interview, both Gugu and Lindi also displayed the same kind of conflicting cultural models.

> Gugu: To me those different languages must be respected; we must never look down upon different people speaking different languages. I think to me they are all important. Much as we are respecting English as an international language, but I think it is high time that we realize that we need to interact with other languages.

While Gugu wants to respect and honor the African languages, she still feels pressured by the international nature of English. During the pre-observation interview, she was emphatic about the need to focus on English; and in the focus group interview, she emphasizes the need to respect and interact with other languages.

The preceding finding highlight the teachers' preference for English as the language of learning and teaching Mathematics and the cultural models that inform the preference. The discussion also shows the conflicting cultural models with which teachers work. A glaring absence in the teachers' cultural models is any reference to how learning and teaching in English, as they prefer, would create epistemological access for the learners. This absence suggests that the teachers position themselves in relation to English and not Mathematics. What are more prevalent in the reasons for preference of English are economic, political and ideological factors. The section that follows explores the learners' language preferences and how they relate to those of teachers.

LEARNERS' LANGUAGE CHOICES

The data used here are drawn from a wider that involved secondary school learners. We analyze individual interviews with five Grade 11 (16-year-old) learners from Soweto, the largest and most multilingual African township in South Africa with a population of about three million people. All of these learners are multilingual (they speak four or more languages) and learn Mathematics in English, which is not their home language. They chose their preferred language for the interview. With the exception of one (Basani), all their schooling has been in Soweto. They all made a choice to do Mathematics and indicated that they like doing Mathematics. Three indicated that they prefer to be taught Mathematics in English while the other two felt that it really does not matter in what language Mathematics is learned.

For the learners who preferred to be taught English (Tumi, Sipho and Nhlanhla), the cultural model of *English as an international language*, which positions English as the route to success, emerged as dominant in their discourse. Their preference for English is because of the social goods that come with the ability to communicate in English.

> Tumi: English is an international language; just imagine a class doing maths with Setswana for example, I don't think it's good.
>
> Researcher: Why?
>
> Tumi: I don't think it is a good idea. Let's say she taught us in Setswana, when we meet other students from other schools and we discuss a sum for instance and she is a white person. I only know division in Setswana, so I must divide this by this and don't know English, then he and I going to have problem. So I think we should talk English. English is okay.

Tumi sees English as an obvious language for learning and teaching Mathematics. It is unimaginable to him for Mathematics to be taught in an African language like Setswana. The use of English as a language of learning and teaching Mathematics is common sense to him; he cannot imagine Mathematics without English. This resonates with the teachers' cultural models, which are exacerbated by the fact that Mathematics texts and examinations are in English. Another factor that emerges from Tumi's views is the fact that he wants to be taught Mathematics in English, so that he can be able to talk about Mathematics in English with white people.

> Sipho: I prefer that ba rute ka English gore ke tlo ithuta ho bua English. If you can't speak English, there will be no job you can get. In an interview, o thola hore lekgowa ha le kgone ho bua Sesotho or IsiZulu, ha o sa tsebe English o tlo luza job. *(I prefer that they teach us in English so that I can learn English. If you can't speak English, there will be no job you can get. In an interview you will find a white person not able to speak Sesotho or IsiZulu, you will loose the job because you don't know English.)*

Sipho's preference for English is because he sees it as a language that gives access to employment. He also connects employment with White people by arguing that during an interview, one must be able to express oneself in English because White people conduct interviews. This connection of jobs to white people and English is a result of the socio-political history of South Africa in which the economy was and still continues to be in the hands of White people, with English as the language of commerce: hence, Sipho's expectation that a job interview will be conducted by a White person in English. Like Gugu, Tumi and Sipho see the Mathematics class as an opportunity for them to gain access to English—the language of power.

Unlike Tumi and Sipho, Nhlanhla, who also indicated a preference for English, positioned herself in relation to Mathematics. Nhlanhla, however, had conflicting cultural models:

> Nhlanhla: . . . is the way it is supposed to be because English is the standardized and international language.

Researcher: Okay, if you had a choice in what language would you choose to learn Mathematics?

Nhlanhla: For the sake of understanding it, I would choose my language. But I wouldn't like that [English as language of learning and teaching] to be changed because somewhere somehow you would not understand what the word 'transpose' mean, ukhithi uchinchela ngale *(that you change to the other side)*; some people won't understand. They would not understand what it means to change the sign and change the whole equation.

While Nhlanhla recognizes the value of learning Mathematics in a language that she understands better, she does not want English as LoLT to change because English is international and the African languages do not have a well-developed Mathematics register. There are conflicting cultural models at play here: one that values the use of African languages for mathematical understanding and another that values English because of its international nature.

Researcher: What if there are students who want to learn Mathematics in Zulu, what would you say to them?

Nhlanhla: I would say its okay to have it but you have to minimize it because these days everything is done in English, especially Maths, Physics and Biology.

Researcher: Why do Maths, Physics and Biology have to be done in English?

Nhlanhla: I don't know; think that's the way it is.

Nhlanhla's conflicting cultural models are evident in the preceding extract. They are indicative of the multiple identities that she is enacting. As a multilingual learner who is not fully proficient in English, she does not want to loose the social goods that come with English. As a Mathematics learner, it is important for her that she has a good understanding of Mathematics and using her language, as she says, facilitates understanding. While the teachers (Kuki, Lindi and Gugu) also experienced conflicting cultural models, theirs were about access to social goods and not to Mathematics.

Basani and Lehlohonolo are the two learners who felt that it really does not matter what language is used for Mathematics. As indicated earlier, Basani is new in the school. Before coming to the school in Soweto, he was a student at a suburban school, which was formerly for Whites only. At the time of the study, it was his second year at the Soweto school to which he came because his mother could no longer afford the fees at the former White school. Basani's level of English fluency was clearly above all the other learners interviewed. During the interview, he explained that he was doing Grade 11 for the second time because he failed IsiZulu and Mathematics the previous year. He, nevertheless, insisted that he has no problem with

Mathematics and that he failed Mathematics because he was not as focused as he should have been.

> Basani: Maths is also a language on its own; it doesn't matter what language you teaching it. It depends if the person is willing to do it.
>
> Researcher: What would you say to learners who want to be taught Mathematics in their African languages?
>
> Basani: I would not have problem. If that's the way they wanna do it, well its their choice. I have a friend here at school; he is Sotho. I help him with Maths. Sometimes when I explain in Sesotho, he doesn't understand; and when I explain it in English, he understands.

As the preceding excerpt shows, Basani believes that Mathematics is a language and, thus, it does not make any difference in what language it is taught and learned. Basani is very confident about his mathematical knowledge and seems to be working with a cultural model that says the key to Mathematics learning is the willingness to do it. Lehlohonolo, who is also very confident about his mathematical knowledge, also felt that it does not matter what language is used for Mathematics. The class teacher explained that he is the best performing learner in Mathematics in his class. Another interesting thing is that when we gave them the information letters and consent forms to participate in the study, Lehlohonolo immediately indicated that we should use his real name because he wants to be famous. During the interview, Lehlohonolo focused more on Mathematics rather than language.

> Lehlohonolo: To me it doesn't matter just as long as I am able to think in all languages and I can speak and write in those languages, then I can do maths in those languages.

Lehlohonolo is connecting language to learning in very sophisticated ways. For him, fluency in a language (ability to read, speak, write and think) facilitates ability to learn in the language. As he explains, fluency in a language is not sufficient to make a learner successful in Mathematics. During the interview Lehlohonolo further argued that language cannot be blamed for failure or given credit for success in Mathematics. He sees the important factor in succeeding in Mathematics as being the learners themselves and the choices they make about how they participate in the Mathematics class. The preceding excerpt suggests that Lehlohonolo enacts a cultural model that Mathematics should be taken only by those who are good at it, and being good at Mathematics is not connected to language.

> Researcher: So if you had a group of students who want to do Mathematics in Zulu, what would you say to them?

> Lehlohonolo: That's their own problem because if they out of high school, they cannot expect to find an Indian lecturer teaching maths in Zulu. English is the simplest language that everyone can speak, so they will have to get used to English whilst they are still here.

While Lehlohonolo does not connect failure or success in Mathematics to language, in the preceding extract he seems to be suggesting that learners should choose to learn in English because in higher education no lecturer will be able to teach in their languages. This is an emergence of a conflicting cultural model for Lehlohonolo, which says even if there is no causal link between success in Mathematics and the language used for learning and teaching, English cannot be ignored.

The preceding discussion shows that the learners' positioning and cultural models are not as clear as those of teachers are. What one can see is that the learners who prefer to be taught in English position themselves in relation to English. Nhlanhla is the only one who preferred English, but she also positioned herself in relation to Mathematics. Tumi and Sipho are more concerned with gaining fluency in English, so that they can access social goods such as jobs and higher education. They enact the same cultural model as teachers that *English is international.* This cultural model emphasizes the belief that the acquisition of the English language constitutes the major content of schooling. This is inconsistent with the content of schooling, which is about giving epistemological access to research and the Language in Education Policy (LiEP) in South Africa, which promotes multilingualism and encourages the use of the learners' home languages. The assumption embedded in this policy is that Mathematics teachers and learners in multilingual classrooms together with their parents are somehow free of economic, political and ideological constraints and pressures when they apparently freely opt for English as LoLT. The LiEP seems to be taking a structuralist and positivist view of language, one that suggests that all languages can be free of cultural and political influences.

As indicated earlier, the learners who position themselves in relation to Mathematics seem to be working with conflicting cultural models—one that is about mathematical understanding and the other that is about English fluency. While teachers also worked with conflicting cultural models, they did not position themselves in relation to Mathematics.

WHAT DOES THIS MEAN FOR RESEARCH AND PRACTICE?

Some researchers argue that to facilitate multilingual learners' participation and success in Mathematics, teachers should recognise their home languages as legitimate languages of mathematical communication (Khisty, 1995; Moschkovich, 1999 & 2002; Setati & Adler, 2002). As alluded to earlier, all the studies that recommend the use of the learners' home languages have been framed by a conception of mediated learning, where language is seen as a tool for thinking and communicating. The studies foreground Mathematics, but the do not consider the political role of language. The analysis presented here shows that the language choices of teachers and learners who prefer English are informed by the political nature of language. The challenge is in bringing the two together. Research shows that in bringing the two together, English dominates.

A recent detailed analysis of a lesson taught by Kuki suggests a relationship between the language(s) used, Mathematics discourses, and the cultural models that emerged (Setati, 2005). During the lesson, Kuki switched between English and Setswana. However, her use of English tended to produce procedural discourse while her use of Setswana tended to produce conceptual discourse. The same observations were made in Lindi's class, as procedural discourse was dominant in Gugu's class who used only English during her teaching. While it can be argued that the observations made in Kuki, Lindi and Gugu's classrooms cannot be generalized to all the teachers in multilingual classrooms, they give us an idea of what the dominance of English in multilingual Mathematics classrooms can produce.

Recent research in South Africa points to the fact that procedural teaching is dominant in most multilingual classrooms (Taylor and Vinjevold, 1999). In most cases, this dominance of procedural teaching is seen as a function of the teachers' lack of or limited knowledge of Mathematics. What the preceding discussion suggests is that the problem is much more complex.

In sum, the preceding analysis shows that teachers and learners who position themselves in relation to English are concerned with access to social goods and are positioned by the social and economic power of English. They argue for English as LoLT. Issues of epistemological access are absent in their discourse. In contrast, learners who position themselves in relation to Mathematics and so epistemological access, reflect more contradictory discourses, including support for the use of the learners' home languages as LoLT. The findings presented here provide an important contribution in dealing with the complex issues related to teaching and learning in multilingual classrooms. Indeed, much remains to be done!

A RESPONSE TO ARTHUR POWELL: DOES ETHNOMATHEMATICS = MATHEMATICS = ANTIRACISM?

As stated earlier, this section is a response to Arthur Powell's paper entitled "Ethnomathematics and the Challenges of Racism in Mathematics Education" (2002). In our view, Powell's paper does not go beyond advocacy for

Ethnomathematics. The paper has three parts. First, he uses the *Adamastor* episode as a metaphor for a racial and political dimension of Ethnomathematics. He does this in order to explain what Ethnomathematics is and its political role in the teaching and learning of Mathematics. Second, he gives an exemplar of how Ethnomathematics can be used in the teaching and learning of Algebra. Here, he focuses on the Ahmose Mathematics papyrus. Third, he moves on to ask the following question: "Who is learning and who is not even in the classroom?" Here, he focuses specifically on small-group learning, which he seems to suggest is consistent with Ethnomathematics and antiracist teaching practices.

Here, we challenges Powell's definition of Ethnomathematics and discuss the implications of constructing Ethnomathematics as a discipline/institution separate and different from Mathematics. Furthermore, we argue that there exists no relationship between Ethnomathematics, small group work, and antiracist teaching.

Does Ethnomatics = Mathematics? In his paper, Powell uses D'Ambrosio's (2001) definition of Ethnomathematics:

> Ethnomathematics is the Mathematics practiced by cultural groups such as urban and rural communities, labour groups, professional classes, children of a certain age bracket, indigenous societies, and many other groups that identify themselves through objects and traditions common to the groups (in Powell, 2002: 17).

Furthermore, Powell theorizes that in Ethnomathematics, the prefix "ethno" not only refers to a specific ethnic, national, or racial group, gender, or even professional group but also to a cultural group defined by a philosophical and ideological perspective. Given Powell's theorization, Ethnomathematics can be defined as the Mathematics practiced by a cultural group defined by a philosophical and ideological perspective. The question here, therefore, is this: How different is this from Mathematics? In our view, Mathematics is also practiced by a cultural group defined by a philosophical and ideological perspective. According to Norman Fairclough (1995), institutions (such as Ethnomathematics or Mathematics) construct their philosophical, ideological and discoursal subjects in a sense that they impose philosophical, ideological and discoursal constraints upon them as a condition for qualifying them to act as subjects. For example, to be a mathematician, one is expected to master the philosophical (ways of being), ideological (ways of seeing) and discursive (ways of talking) norms which the Mathematics community attaches to that position. These ways of being, seeing and talking are inseparably intertwined in the sense that in the process of acquiring the ways of being which are associated with a subject position, one necessarily also acquires its ways of seeing and talking. But, of course, this is not a one-way process. Mathematicians talk and act the way they do because they are mathematicians, but being a mathematician is what it is because there are people who talk and act in the way mathematicians do. The 'Mathematics world' both pre-exists and shapes how mathematicians talk and act in and on it, and it means what it means and has the shape it does because Mathematics educators talk and act in and on it as they do. The same can be said about Ethnomathematics.

The question to ask here is this: Are the philosophical, ideological and discursive norms of Ethnomathematics different from those of Mathematics? Renuka Vithal and Ole Skovsmose have argued that "Almost all definitions or descriptions of 'Ethnomathematics' feature the term Mathematics in various ways as 'mathematical knowledge', 'mathematical ideas', 'mathematical activities' or 'mathematical practices'" (1997:13). Powell's definition of Ethnomathematics also suggests that Ethnomathematics is a special type of Mathematics. What does it mean, therefore, to construct Ethnomathematics as an institution separate and different from Mathematics?

There is no doubt that the construction of Ethnomathematics as an independent and different discipline/institution from Mathematics has created an opportunity for Ethnomathematics researchers to organize and mobilize themselves. It has also led to a creation of a social identity for all the researchers involved in Ethnomathematics research. There are now international conferences and publications focusing on Ethnomathematics. This has raised awareness of the contribution that non-European communities have made to Mathematics. Naming, however, creates boundaries and emphasises difference and, thus, can be counter-productive. Naming in this case has created the perception that Ethnomathematics is different from Mathematics and thus inferior. In my view, Ethnomathematics is Mathematics and, therefore, to construct it as separate from Mathematics is to marginalize it. It is important that Ethnomathematics moves from the margins into the centre of Mathematics.

Does Ethnomathematics = Antiracism? Powell argues that Ethnomathematics has a political focus; it is imbued with ethics and focused on the recuperation of cultural dignity of human beings. We agree with Powell that Ethnomathematics challenges the universal conception of Mathematics knowledge that privileges the dominant groups. Powell's example of how he uses Ethnomathematics in his teaching assumes that bringing cultural contexts into Mathematics classrooms is unproblematic and in fact reconciles differences and affirms learners from marginalized communities. What Powell is not considering is that bringing the learners' cultural backgrounds into the classroom has a potential to reproduce the inequalities that exist in the broader society. It is therefore important to ask whether there has been any confirmation that knowledge of Ethnomathematics results in action against oppression and domination.

In his paper, Powell argues that Ethnomathematics has a potential of ensuring that the 'African' is not only present but also learning in a Mathematics classroom. In his elaboration of this argument, Powell suggests small-group discussion as a useful pedagogy. While we agree with Powell about the merits of small-group work in Mathematics teaching, it is not clear from Powell's paper what small-group work has to do with Ethnomathematics or antiracist teaching and learning practices. Our questions to Powell here are the following: Is there a relationship between small-group learning and Ethnomathematics? Is there a relationship between antiracist teaching and practices and Ethnomathematics? From our experiences as Mathematics learners, educators and researchers, there is antiracist teaching and learning practices taking place in many non-Ethnomathematics classrooms.

In conclusion, we believe that it is time that the Ethnomathematics community re-evaluates its agenda. Ethnomathematics emerged, in part, as a consciousness movement within Mathematics education, critiquing the conception of Mathematics knowledge that privileges the dominant groups. There is also a need for Ethnomathematics to turn the critique inward (Vithal and Skovsmose, 1997: 22).

Bibliography

Abraham, R. 1967. *The Principles of Ibo*. Ibadan, Nigeria: University of Ibadan Press.

Abraham, R. H. 1996. Chaos and the millennium. A Talk for the National Collegiate Honors Council, San Francisco, California, October 31, 1996. Visual Math Institute (http://www.vismath.org). Retrieved on August 8, 2008. Available at http://www.gaianxaop.com/xaos_millennium.htm

Acheson, D. J. 2005. *Elementary Fluid Dynamics*. New York, NY: Oxford University Press.

Adendorff, R. 1993. Code-switching amongst Zulu-speaking teachers and their pupils. *Language and Education* 7, 3:141–162.

Adler, P. and N. Barnard. 1992. *African Majesty: The Textile Art of the Ashanti and Ewe*. London, UK: Thames and Hudson.

Adler, J. 2001. *Teaching Mathematics in Multiligual Classrooms*. Dordrecht, The Netherlands: Kluwer Academic Press.

Adler, J. 1994. Mathematics teachers in the South African transition. *Mathematics Education Research Journal* 6, 2:101–112.

African-American Institute. 1976. The new American mathematics and the old African mathematics: An adventure in comparative mathematics. *Mini-Module*. New York, NY: African-American Institute Publications.

Agbinya, J. I. 2004. *Computer Board Games of Africa: Algorithms, Strategies and Rules*. Bellville, South Africa: Department of Computer Science, University of Western Cape.

Alshameri, F. J. 2006. Automated generation of metadata for mining image and text data. Doctoral dissertation, George Mason University, Fairfax, Virginia.

Altschiller-Court, N. 1964. The dawn of demonstrative geometry. *Mathematics Teacher* 57:163–166.

Arfken, G. B. and H. J. Weber. 2005. *Mathematical Methods for Physicists 6th ed.* Amsterdam, The Netherlands: Elsevier Academic Press.

Arthur, J. 1994. English in Botswana primary classrooms: Functions and constraints. In C. M. Rubagumya, ed. *Teaching and Researching Language in African Classrooms*. Clevedon, UK: Multilingual Matters Ltd.

Arnold, V. I. 1992. *Catastrophe Theory 3rd ed.* Berlin, Germany: Spirnger-Verlag.

Ascher, M. 1988. Graphs in cultures. II: A study in ethnomathematics. *Archive for History of Exact Sciences* 39, 1:75–95.

Ascher, M and R. Ascher. 1986. Ethnomathematics. *History of Science* 24, 64, part 2:125–144.

Aveni, A. F. 1981. Tropical archeoastronomy. *Science* 213, 4504:161–171.

Baker, G. L. and J. P. Gollub. 1990. *Chaotic Dynamics*. Cambridge: Cambridge University Press.

Bangura, A. K. 2005. Ubuntugogy: An African educational paradigm that transcends pedagogy, andragogy, ergonagy and heutagogy. *Journal of Third World Studies* xxii, 2:13–54.

Bangura, A. K. 2001. Book Review of Ron Eglash's *African Fractals: Modern Computing and Indigenous Design. Nexus Network Journal* 2, 4.

Bangura, A. K. 2000. Book Review of George Gheverghese Joseph's *The Crest of the Peacock: No-European Roots of Mathematics 2nd ed. Nexus Network Journal* 3, 3.

Bangura, A. K. 2000. *Chaos Theory and African Fractals*. Washington, DC: The African Institution.

Barker, J. and J. Thornton. 2004. Software engineering challenges in bioinformatics. *Proceedings of the 26th International Conference on Software Engineering* (ICSE '04).

Barnsley, M. F. 1993. *Fractals Everywhere* 2nd ed. New York: Ap Professional.

Bastin, M.-L. 1961. *Art Décoratif Tshokwe*. Lisbon, Portugal: Publiçaões Culturais da Companhia de Diamantes de Angolas.

Berry, M., Z. Drmac and E. Jessup. 1999. Matrices, vector spaces, and information retrieval. *Society for Industrial Applied Mathematics* (SIAM) 41, 2:335–362.

Biggs, N. L. 1979. The roots of combinatorics. *Historia Mathematica* 6, 2:109–136.

Bogoshi, J., K. Naidoo and J. Webb. 1987. The oldest mathematical artifact. *Math Gazette* 71:458.

Bohensky, E. L., B. Reyers and A. S. Van Jaarsveld. 2006. Future ecosystem services in a Southern Africa river basin: A scenario planning approach to uncertainty. *Conservation Biology* 20, 4:1051–1061.

Bolland, R. 1991. *Tellem Textiles: Archaeological Finds from Buried caves in Mali's Bandiagara Cliff.* Bamako, Mali: Musée National/Amsterdam, Holland: Tropenmuseum.

Bracewell, R. N. 1999. *The Fourier Transform and Its Applications. 3rd ed.* New York, NY: McGraw-Hill Book Company.

Briggs, J. 1992. *Fractals: The Patterns of Chaos: A New Aesthetic of Art, Science, and Nature*. New York: Touchstone Books.

Briggs, K. 1997. Feigenbaum scaling in discrete dynamical systems. Ph.D. Thesis, University of Melbourne, Australia.

Briggs, N. L. The Roots of Combinatorics. *Historia Mathematica* 6, 2:109–136.

Brigham, E. O. 1988. *The Fast Fourier Transform and Applications*. Englewood Cliffs, NJ: Prentice Hall.

Broer, H. et al. 2003. *Bifurcations in Hamiltonian Systems: Computing Singularities by Gröbner Bases*. New York, NY: Springer-Verlag.

Broline, D. M. and D. E. Loeb. 2008. The Combinatorics of Mancala-type games: Ayo, Tchoukaillon, and 1/ϖ. *arXiv:Math/9502225v*1:1–14. [math.CO, February 9, 1995].

Brown, C. 1995. *Chaos and Catastrophe Theories*. Thousand Oaks, CA: Sage Publications.

Brualdi, R. A. 1992. *Introductory Combinatorics*. New York, NY: North-Holland Publishing Company.

Bruins, E. M. 1981. Egyptian arithmetic. *Janus* 68, 1–3:33–52.

Brunner, R., S. Djorgovsky, T. Prince and A. Szalay. 2002. Massive datasets in astronomy. In J. Abello et al., eds. *Handbook of Massive Datasets*. Norwell, MA: Kluwer Academic Publishers.

Brusco, M. J. and S. Stahl. 2005. *Branch-and-Bound Applications in Combinatorial Data Analysis* (*Statistics and Computing* Series). New York, NY: Springer.

Buffoni, B and J. Toland. 2003. *Analytic Theory of Global Bifurcation*. Princeton, NJ: Princeton University Press.

Cannataro, M., C. Comito, A. Guzzo and P. Veltri. 2004. Integrating ontology and workflow in PROTEUA, a grid-based problem solving environment for bioinformatics. *Proceedings of the International Conference on Information Technology: Coding and Computing* (ITCC '04).

Chen, M., J. Han and P. Yu. 1996. Data mining: An overview from a database perspective. *IEEE Transactions on Knowledge and Data Engineering* 8, 6:866–883.

Ciucu, M. November 2005. *A Random Tiling Model for Two Dimensional Electrostatistics* (Memoirs of the American Mathematical Society, vol. 178, no. 839). Providence, RI: American Mathematical Society.

Cohen, A. and R. D. Ryan. 1995. *Wavelets and Multiscale Signal Processing*. London, UK: Chapman and Hall.

Constantine, G. M. 1987. *Combinatorial Theory and Statistical Design*. New York, NY: John Wiley and Sons, Inc.

Corwin, L. J. and R. H. Szczarba. 1995. *Calculus in Vector Spaces 2nd ed.* New York, NY: Marcel Dekker, Inc.

Crawford, J. D. 1991. Introduction to bifurcation theory. *Reviews of Modern Physics* 63, 4:991–1037.

Crowe, M. J. 1994. *A History of Vector Analysis: The Evolution of the Idea of a Vectorial System*. Mineola, NY: Dover Publications.

Crowe, D. W. 1982. The geometry of African art III: The smoking pipes of Begho. In C. Davis et al., eds. *The Geometric Vein, the Coxeter Festchrift*. New York, NY: Springer Verlag.

Crowe, D. W. 1975. The geometry of African art II: A catalog of Benin patterns. *Historia Mathematica* 2:253–271.

Crowe, D. W. 1971. The geometry of African art I: Bakuba art. *Journal of Geometry* 1:169–182.

Cummins, J. 2000. *Language, Power and Pedagogy: Bilingual Children in the Crossfire*. Clevedon, UK: Multilingual Matters Ltd.

Davis, P. J. 1993. *Spirals: From Theodorus to Chaos*. Wellesley, MA: A. K. Peters.

de Heinzelin de Braucourt, J. June 1962. de Ishango. *Scientific American* 206,6:105–116.

Delafosse, M. 1929. *La Langue Mandingue et ses Dialectes*, vol. 1. Paris, France: Librarie Orientaliste, Paul Guenther.

Demazure, M. 2000. *Bifurcations and Catastrophes: Geometry of Solutions to Nonlinear Problems*. New York, NY: Springer-Verlag.

Devaney, R. L. 1992. *A First Course in Chaotic Dynamical Systems*. New York: Addison-Wesley Publishing Co.

Devaney, R. L. 1989. *Chaos, Fractals, and Dynamics: Computer Experiments in Mathematics*. New York: Addison-Wesley.

Dhillon, I., J. Han and Y. Guan. 2001. Efficient clustering of very large document collection. In R. Grossman et al., eds. *Data Mining for Scientific and Engineering Applications*. Norwell, MA: Kluwer Academic Publishers.

Dike, O. A. W. 1987. *Mathematics and Measurement: Reading the Past.* Berkeley, CA: University of California Press.

Diop, C. A. 1980. *Civilisation ou Barbarie: Anthrologies sans Complainsance*. Paris, France: Présence Africaine.

Djebbar, A. 1995. On mathematical activities in North Africa since the 9th Century. *AMUCHA Newsletter 15*. Retrieved on January 25, 2007 from http://www.math.buffalo.edu/mad/AMU/amu+chma_15.html#2

Donkers, J., J. Uiterwijk and A. de Voogt. 2001. Mancala games—topics in mathematics and artificial intelligence. *Board Games in Academia* IV, April 17–21, 2001. Fribourg, Switzerland: University of Fribourg.

Dorre, J., P. Gerstl and R. Seiffert. 1999. Text mining: Finding nuggets in mountains of textual data. KDD-99:398–401. San Diego, CA.

Dumortier, F. et al. 1991. *Bifurcations of Planar Vector Fields: Nilpotent Singularities and Abelian Integrals*. New York, NY: Springer-Verlag.

Eglash, R. in press. From hip-hop to flip-flop: Black noise in the master-slave circuit. In B. Sinclair and R. Fouche, eds. *Technology and the African American Experience: Needs and Opportunities for Study*. Washington, DC: Smithsonian Institution Press.

Eglash, R. 1999. *African Fractals: Modern Computing and Indigenous Design*. New Brunswick, NJ: Rutgers University Press.

Eglash, R. 1998. Fractals in African settlement architecture. *Complexity* November-December, 4, 2.

Eglash, R. 1998. Geometric algorithms in Mangbetu design. *Mathematics Teacher* 91, 5:376–381.

Eglash, R. 1997a. Bamana sand divination: Recursion in ethno-mathematics. *American Anthropologist* 99, 1:112–122.

Eglash, R. 1997b. The African heritage of Benjamin Banneker. *Social Studies of Science* April.

Eglash, R. 1997c. When math worlds collide: Intention and invention in ethnomathematics. *Science, Technology and Human Values* Winter, 22, 1:79–97.

Eglash, R., E. Onyejekwe, C. S. Diatta and N. Badiane. 1997. Indigenous science for education and development: A boot-strapping approach. *African Fractals in Development*. Retrieved on June 27, 2008. Available at http://www.rpi.edu/~eglash/eglash.dir/develop.htm

Eglash, R. 1995a. Fractal geometry in African material culture. *Symmetry: Culture and Science* 6, 1:174–177.

Eglash, R. 1995b. Scaling hexagons in a Bassari initiation mask. *Mathematics Teacher* October, 618–620.

Eglash, R. 1995c. African influences in cybernetics. In C. Gray, ed. *The Cyborg Handbook*. New York: Routeledge.

Eglash, R. 1995d. African influence in cybernetics. Retrieved on June 27, 2008. Available at http://www.haussite.net/haus.0/SCRIPT/txt2001/01/eglash.HTML

Eglash, R. 1995e. Fractal geometry in African material culture (Symmetry: Natural and Artificial Extended). *Symmetry: Culture and Science* 6, 1:174–177.

Eglash, R. and P. Broadwell. 1995. Fractals in architecture. In B. Wahl, ed. *Exploring Fractals on the Macintosh*. New York: Addison-Wesley.

Eglash, R., C. Diatta and N. Badiane. 1994. Fractal structure in jola material culture. *Ekistics* September-December, 61, 368/369:367–371.

Eglash, R. and P. Broadwell. 1989. Fractal geometry in traditional African architecture. *Dynamics Newsletter* June.

Engels, H. 1977. Quadrature of the circle in ancient Egypt. *Historia Mathematica* 4:137–140.

Erdös, P. et al. 1984. *Combinatorial Set Theory: Partition Relations for Credentials* (Studies in Logic and the Foundations of Mathematics Series No. 106). Amsterdam, The Netherlands: North-Holland Publishing Company.

Erdös, P. and R. Rado. 1956. "A Partition Calculus in Set Theory." *Bulletin of the American Mathematical Society* 62:427–489.

Erickson, M. J. 1996. *Introduction to Combinatorics*. New York, NY: John Wiley and Sons, Inc.

Ester, M., H. Kriegel and J. Sander. 2001. Algorithms and applications for spatial data mining. *Geographic Data Mining and Knowledge Discovery, Research Monographs in GIS*. Taylor and Francis, 167–187.

Ester, M., H. Kriegel and J. Sander. 1997. Spatial data mining: A database approach. *Proceedings of the International Symposium on Large Spatial Databases*. SSD '97:47–66. Berlin, Germany.

Ester, M., A. Frommelt, H. Kriegel and J. Sander. 2000. Spatial data mining: Database primitives, algorithms and efficient DBMS support. *Data Mining and Knowledge Discovery* 4, 2/3:193–216.

Evans, B. 1993. Number and form and content: A composer's path to inquiry. In Leonardo Book Series, *The Visual Mind*. Cambridge, MA: MIT Press.

Eves, H. 1958. On the practicality of the Rule of False Position. *Mathematics Teacher* 51:606–608.

Field, M. and M. Golubitsky. 1995. Symmetric chaos. *Math Horizons* April, 34–35.

Fairclough, N. 1995. *Critical Discourse Analysis: The Critical Study of Language*. London, UK: Longman.

Farum-Badley, W. L. n.d. Barbados Warri: The Bajan way. Retrieved on July 27, 2008. Available at http://barbadosphotogallery.com/warri/warrigame.htm

Fauvel, J. and P. Gerdes. 1990. African slave and calculating prodigy: Bicentenary of the death of Thomas Fuller. *Historia Mathematica* 17, 2:141–151.

Fayyad U. M., G. Piatetsky-Shapiro and P. Smyth. 1996. From data mining to knowledge discovery: An overview. U. M. Fayyad et al., eds. *Advances in Knowledge Discovery and Data Mining*. Menlo Park, CA: AAAI Press.

Ferrari, M. J. et al. 2008. The dynamics of measles in Sub-Saharan Africa. *Nature* 451:679–684.

Field, M. and M. Golubitsky. 1995. Symmetric chaos: How and why? *Notices* of the American Mathematical Society 42, 2:240–244.

Field, M. and M. Golubitsky. 1992. *Symmetry in Chaos*. Oxford: Oxford University Press.

Fischer, I. K. 1981. At the dawn of geodesy. *Bulletin Géodésique* 55, 2:132–142.

Fletcher, E. N. R. 1970. The area of the curved surface of a hemisphere in ancient Egypt. *Mathematics Gazette* 389:227–229.

Floyd, B. P. 1999. Now important in computer modeling, fractals have a rich history in African design. *The Chronicle of Higher Education* June 11, A31.

Folland, G. B. 1999. *Real Analysis: Modern Techniques and their Applications, 2nd ed.* New York: Wiley.

Folland, G. B. 1992. *Fourier Analysis and Its Applications*. Pacific Grove, CA: Wadsworth & Brooks/Cole Advanced Books & Software.

Françoise, J.-P. and R. Roussarie, eds. 1990. *Bifurcations of Planar Vector Fields: Proceedings of a Meeting Held in Luminy, France, September 18–22, 1989*. New York, NY: Springer-Verlag.

Garcia, J, B. 1997. Number systems and calendars of the Berber populations of Grand Canary and Tenerife in the 14th-15th Centuries. Doctoral Thesis, University of La Laguna, Tenerife.

Gasquet, C. and P. Witomski. 1999. *Fourier Analysis and Applications: Filtering, Numerical Computation, Wavelets*. New York, NY: Springer-Verlag.

Gee, J. P. 1999. *An Introduction to Discourse Analysis: Theory and Method*. London, UK: Routledge.

Gee, J. P. 1996. *Social Linguistics and Literacies: Ideology in Discourses*. London, UK: Falmer Press, a member of the Taylor and Francis Group.

Gerdes, P. 2007a. *Sona Geometry from Angola: Mathematics of an African Tradition*. Milan, Italy: Polimetrica, International Scientific Publisher.

Gerdes, P. 2007b. *Drawings from Angola: Living Mathematics*. US, UK, Canada: Lulu.com.

Gerdes, P. 2007c. *African Doctorates in Mathematics: A Catalogue*. US, UK, Canada: Lulu.com.

Gerdes, P. and A. Djebbar. 2007. *Mathematics in African History and Cultures: An Annotated Bibliography*. US, UK, Canada: Lulu.com.

Gerdes, P. 1999. *Geometry from Africa: Mathematical and Educational Explorations*. Washington, DC: The Mathematical Association of America.
Gerdes, P. 1997. *Lusona: Recreation geometriques d'Afrique (Geometrical Recreations of Africa)*. Paris, France: L'Harmattan.
Gerdes, P. 1996a. *Luanda Geometry: Designs, Polyminoes, Patterns, Symmetries*. Maputo, Mozambique: Ethnomathematics Research Project, Instituto Superior Pedagogico.
Gerdes, P. 1996b. *Luanda Sipatsi: Technology, Art, and Geometry in Inhambane*. Maputo, Mozambique: Ethnomathematics Research Project, Instituto Superior Pedagogico.
Gerdes, P. 1994a. *African Pythagoras: A Study in Culture and Mathematics Education*. Maputo, Mozambique: Ethnomathematics Research Project, Instituto Superior Pedagogico.
Gerdes, P. 1994b. On mathematics in the history of Sub-Saharan Africa. *Historia Mathematica* 21, 3:345–376.
Gerdes, P. 1992. Fivefold symmetry and (basket) weaving in various cultures. In *Fivefold Symmetry*. River Edge, NJ: World Science Publishing.
Gerdes, P. 1990. On ethnomathematical research and symmetry: Symmetry in a Kaleidoscope Extended 2. *Symmetry: Culture and Science* 1, 2:154–170.
Gerdes, P. 1985. Three alternate methods of obtaining the ancient Egyptian formula for the area of a circle. *Historia Mathematica* 12, 3:261–268.
Gillings, R. J. 1982. *Mathematics in the Time of the Pharaohs*. New York, NY: Dover Press.
Gillings, R. J. 1972. *Mathematics in the Time of the Pharaohs*. Cambridge, MA: Massachusetts Institute of Technology Press.
Gillings, R. J. 1964. The volume of a truncated pyramid in ancient Egyptian papyri. *Mathematics Teacher* 57:552–555.
Gillings, R. J. 1962. Problems 1 to 6 of the Rhind Mathematical Papyrus. *Mathematics Teacher* 56:61–69.
Gilmer, G. December 31, 1998. Mathematical patterns in African American hairstyles. Paper presented at the 77th Annual Meeting of the National Council of Teachers of Mathematics in San Francisco, California.
Gilmore, R. 1993. *Catastrophe Theory for Scientists and Engineers*. New York, NY: Dover.
Ginsburg, H. 1978. Poor children, African mathematics, and the problem of schooling. *Educational Research Quarterly* 2, 4:26–44.
Glendinning, P. 1994. *Stability, Instability and Chaos*. Cambridge, UK: Cambridge University Press.
Gnaedinger, F. 2005. Very early calendars. Retrieved on July 22, 2008. Available at http:// http://www.seshat.ch/home/calendar.htm
Godsil, C. D. 1993. *Algebraic Combinatorics*. New York, NY: Chapman and Hall.
Golomb, S. 1994. *Polyominoes*. Princeton, NJ: Princeton University Press.
Gomez, M., A. Gelbuhk, A. Lopez and R. Yates. 2001. Text mining with conceptual graphs. *IEEE* 893–903.

Graham, R. L., M. Groeschel and L. Lovász, eds. 1996. *Handbook of Combinatorics* vols. 1 & 2. Cambridge, MA: The MIT Press.

Graham, R. L., et al. 1990. *Ramsey Theory*. New York, NY: John Wiley and Sons, Inc.

Granville, S. et al. 1998. English without g(u)ilt: A position paper on language in education policy for South Africa. *Language in Education* 12, 4:254–272.

Griffiths, D. J. 1999. *Introduction to Electromagnetics 3rd ed.* San Francisco, CA: The Benjamin/Cummings Company.

Grossman, R., E. Creel, M. Mazzucco and R. Williams. 2001. A dataspace infrastructure for astronomical data. In R. Grossman et al., eds. *Data Mining for Scientific and Engineering Applications*. Norwell, MA: Kluwer Academic Publishers.

Grünbaum, B. 1984. The emperor's new clothes: Full regalia, G-string, or nothing? *Mathematics Intelligencer* 6, 4:47–56.

Grünbaum, B. and G. C. Shephard. 1987. *Tilings and Patterns*. New York, NY: W. H. Freeman and Company.

Gulick. D. 1992. *Encounters With Chaos*. New York: McGraw Hill College Division.

Gutiérrez, R. 2002. Enabling the practice of mathematics teachers in context: Toward a new equity research agenda. *Mathematical Thinking and Learning* 4, 2:145–187.

Hall, S. G. 1999. Towards a working non-linear science of empowerment. Paper presented at the Ninth Annual International Conference of the Society for Chaos Theory in Psychology and the Life Sciences, Berkeley, California, July 23–26–1999.

Hall, Jr., M. 1986. *Combinatorial Theory*. New York, NY: John Wiley and Sons, Inc.

Halmos, P. R.1964. *Measure Theory*. New York, NY: Van Norstrand Company.

Hambrusch, S., C. Hoffman, M. Bock, S. King and D. Miller. 2003. Massive data: Management, analysis, visualization, and security. A School of Science Focus Area at Purdue University Report.

Han, J. and M. Kamber. 2001. *Data Mining: Concepts and Techniques*. San Francisco, CA: Morgan Kaufman Publishers.

Hargittai, I. and C. A. Pickover, eds. 1992. *Spiral Symmetry*. Singapore: World Scientific Publishing Co.

Harsthorne, K. B. 1987. Language policy in Africa education in South Africa, 1910–1985, with particular reference to the issue if medium of instruction. In D. Young, ed. *Bridging the Gap: Essays in Honour of L. W. Lanham*. Pretoria, South Africa: Maskew Miller Longman.

Hauenstein, A. 1988. *Examen de Motifs Décoratifs chez les Ovimbundu et Tchokwe d'Angola*. Coimbra, Portugal: Universidade de Coimbra.

Helck, T. n.d. Ten-pointed star patterns: Examples of tessellation in Islamic art. Retrieved on July 11, 2008. Available at http://www.thelck.com/patterns/tenPointDtar.html

Heugh, K., ed. 1997. *Multilingual Education for South Africa-*. Johannesburg, South Africa: Heinemann.

Hildebrand, S. and A. Tromba. 1996. *The Parsimonious Universe: Shape and Form in the Natural World*. New York, NY: Copernicus.

Hoadley, U. 2007. The reproduction of social class inequalities through mathematics pedagogies in South African primary schools. *Journal of Curriculum Studies* 39, 6:679–706.

Holden, A. V., ed. 1986. *Chaos*. Princeton, NJ: Princeton University Press.

Holland, D. and N. Quinn, eds. 1987. *Cultural Models in Language and Thought*. Cambridge, UK: Cambridge University Press.

Howie, S. 2002. *English Language Proficiency and Contextual Factors Influencing Mathematics Achivement of Secondary School Pupils in South Africa*. Den Hagg/ The Hague, The Netherlands: CIP-Gegevens Koninklijke Bibliotheek.

Howie, S. 1997. Mathematics and science performance in the middle school years in South Africa: A summary report on the performance of South African students in the Third International Mathematics and Science Study. Pretoria, South Africa: Human Sciences Research Council.

Howie, S. and C. A. Hughes. 1999. Mathematics and science literacy of final-year school students in South Africa: A report on the performance of South African students in the Third International Mathematics and Science Study. Pretoria, South Africa: Human Sciences Research Council.

Howie, S. and J. J. Pietersen. 1997. Mathematics literacy of final year students: South African realities. Pretoria, South Africa: Human Sciences Research Council.

Howie, S. et al. 2000. Middle school students' performance in mathematics in the Third International Mathematics and Science Study: South African realities. Pretoria, South Africa: Human Sciences Research Council.

Høyrup, J. 1990. Sub-scientific mathematics: Observations on a pre-modern phenomenon. *History of Science* 28, 79, part 1:63–87.

Hubbard, B. B. 1996. *The World According to Wavelets*. Wellesley, MA: A. K. Peters.

Humphreys, R., J. Cabanela and J. Kriessler. 2001. Mining astronomical databases. In R. Grossman et al., eds. *Data Mining for Scientific and Engineering Applications*. Norwell, MA: Kluwer Academic Publishers.

Ilukor, Y. 1978. *The Game of Amwesoro*. Kampala, Uganda: Department of Physics, Makerere University.

Imhausen, A. 2007. Egyptian mathematics. In V. J. Katz. *The Mathematics of Egypt, Mesopotamia, China, India, and Islam: A Sourcebook*. Princeton, NJ: Princeton University Press.

Jackson, D. 1963. *Fourier Series and Orthogonal Polynomials*. Washington, DC: Mathematical Association of America.

Jama, M. A. F. 1983. Evaluation of mathematics curriculum in primary teacher training institute in Somalia. Mogadishu, Somalia: African Studies in Curriculum Development and Evaluation.

James, J. F. 1995. *A Student's Guide to Fourier Transforms with Applications in Physics and Engineering*. New York, NY: Cambridge University Press.

Jennings, G. A. 1994. *Modern Geometry with Applications*. New York: Springer-Verlag.

Jones, P. S. 1980. The history of mathematics—new sources and uses. *Southeast Asian Bulletin of Mathematics* 4, 1:1–5.

Joseph, G. G. 1991/2000. *The Crest of the Peacock: Non-European Roots of Mathematics*. New York, NY: Penguin Books.

Kafatos, M., R. Yang, X. Wang, Z. Li and D. Ziskin. 1998. Information technology implementation for a distributed data system serving earth scientists: Seasonal to International ESIP. *Proceedings of the 10th International Conference on Scientific and Statistical Database Management* 210–215.

Kamath, C. 2001. On mining scientific datasets. In R. Grossman et al., eds. *Data Mining for Scientific and Engineering Applications*. Norwell, MA: Kluwer Academic Publishers.

Kammler, D. W. 2000. *A First Course in Fourier Analysis*. Upper Saddle River, NJ: Prentice Hall.

Kani, A. 1992. African mathematics in the pre-colonial Central Sudan. G. Emeagwali, ed. *Science and Technology in African History*. New York: Edwin Mellen Press.

Katz, V. J., ed. 2007. *The Mathematics of Egypt, Mesopotamia, China, India and Islam: A Sourcebook*. Princeton, NJ: Princeton University Press.

Khisty, L. L. 1995. Maki9ng inequality: Issues of language and meaning in mathematics teaching with Hispanic students. In W. G. Secada, E. Fenema, and L. B. Abajian, eds. *New Directions for Equity in Mathematics Education*. Cambridge, UK: Cambridge University Press.

King, M. and R, Greenstone. 1999. *EOS Reference Handbook*. Washington, DC: NASA Publications.

Koperski, K. and J. Han. 1995. Discovery of spatial association rules in geographic information databases. *Proceedings of the 4th International Symposium on Advances in Spatial Databases*. 47–66. Portland, ME.

Koperski, K. and J. Han and N. Stefanovic. 1998. An efficient two-step method for classification of spatial data. *Proceedings of the Symposium on Spatial Data Handling*. 45–54. Vancouver, Canada.

Körner, T. W. 1993. *Exercises for Fourier Analysis*. Cambridge, UK: Cambridge University Press.

Krishnamurthy, V. 1986. *Combinatorics: Theory and Applications*. West Sussex, UK: Ellis Horwood Limited.

Krajčheski, M. November 2001.*Tilings of the Plane, Hyperbolic Groups and Small Cancellation Conditions* (Memoirs of the American Mathematical Society, vol. 154, no. 733). Providence, RI: American Mathematical Society.

Kunen, K. 1980. *Set Theory: An Introduction to Independence Proofs*. Amsterdam, The Netherlands: North-Holland Publishing Company.

Laszlo, E. 1991. *The Age of Bifurcation: Understanding the Changing World*. Philadelphia, PA: Gordon and Breach.

Laurwerier, H. 1991. *Fractals*. Princeton, NJ: Princeton University Press.

Lawler, E. L. et al., eds. 1985. *The Traveling Salesman Problem: A Guided Tour of Combinatorial Optimization*. New York, NY: John Wiley and Sons, Inc.

Le Roux, R. G. 1985. The South African Mathematics Olympiad: A case study of the medalists from 1966 to 1983. Report 0–181. Pretoria, South Africa: Human Sciences Research Council.

Liebenberg, L. 1990. *The Art of Tracking: The Origin of Science*. Claremont, South Africa: David Philip Publications.
Linder, C. J. and H. T. Hudson. 1989. A comparison of mathematics backgrounds between American and South African physics students. *Science Education* 73, 4:459–465.
Loomis, E. S. 1968/1972. *The Pythagorean Proposition*. Reston, VA: National Council of Teachers of Mathematics.
Lord, E. A. and C. B. Wilson. 1984. *The Mathematical Description of Shape and Form*. New York: Halsted Press.
Lumpkin, B. 1992. From Egypt to Benjamin Banneker: African origins of false position solutions. *Vita Mathematica* 279–289; *MAA Notes* 40, 1996.
Lumpkin, B. 1986. Africa in the mainstream of mathematics history. I. V. Sertima, ed. *Blacks in Science: Ancient and Modern*. New Brunswick, NJ: Transaction Books.
Lumpkin, B. 1980. Note: The Egyptians and Pythagorean triples. *Historia Mathematica* 7, 2:186–187.
Manansala, P. 1995. Sungka mathematics of the Philippines. *Indian Journal of Historical Science* 30, 1:13–29.
Mankiewicz, R. 2001. *The Story of Mathematics*. Princeton, NJ: Princeton University Press.
Marsden, A. and A. J. Tromba. 2003. *Vector Calculus 5th ed*. San Francisco, CA: W. H. Freeman and Company.
Marshak, A. 1972. *The Roots of Civilization*. London, UK: Weidenfeld and Nicholson.
Marusic, I., G. Candler, V. Interrante, P. Subbareddy and A. Moss. 2001. Real time feature extraction for the analysis of turbulent flows. In R. Grossman et al., eds. Data Mining for Scientific and Engineering Applications. Norwell, MA: Kluwer Academic Publishers.
Masota, L. A. 1982. Investigation of the effectiveness of teacher-education curriculum in primary school mathematics in Tanzania mainland. African Studies in Curriculum Development and Evaluation. Nairobi, Kenya: Kenya Institute of Education, African Curriculum Organization.
Mayega, J. V. 1974. *Omweso: A Mathematical Investigation of an African Board Game*. Kampala, Uganda: Department of Mathematics, Makerere University.
Mbiriru, M. W. N. K. 1983. Problems of teaching junior secondary mathematics in Kenya. African Studies in Curriculum Development and Evaluation. Nairobi, Kenya: Kenya Institute of Education, African Curriculum Organization.
McGuire, M. 1991. *An Eye for Fractals*. New York: Addison-Wesley Publishing Co.
Mda, T. November 1997. Learning best in home language. p.6.
Merritt, M. et al. 1992. Socialising multilingualism: Determinants of code-switching in Kenyan primary classrooms. *Journal of Multilingual and Multicultural Development* 13, 1&2:103–121.
Meurant, G. 1986. *Shoowa Design: African Textiles from the Kingdom of Kuba*. London, UK: Thames and Hudson.
Migeod, W. L. 1911. *Languages of West Africa* 2 vols. London, UK: Kegan Paul, Trench, Trubner.

Moore, D. L. 1994. *The African Roots of Mathematics 2nd ed.* Detroit, MI: Professional Educational Services.

Morrison, N. 1994. *Introduction to Fourier Analysis.* New York, NY: Wiley.

Moschkovich, J. 1999. Supporting the participation of English language learners in mathematical discussions. *For the Learning of Mathematics* 19, 1:11–19.

Moschkovich, J. 2002. A situated and sociocultural perspective on bilingual mathematics learners. *Mathematical Thinking and Learning* 4, 2&3:189–212.

Mostafa, M. 2005. Factors affecting organisational creativity and innovativeness in Egyptian business organisations: An empirical investigation. *Journal of Management Development* 24, 1:7–33.

Neugebauer, O. 1980. On the orientation of pyramids. *Centaurus* (special issue dedicated to Olaf Pedersen on his sixtieth birthday) 24:1–3.

Nex, P. A. M. 2004. Formation of bifurcation chromite layers of the UG1 in the Bushveld Igneous Complex, and analogy with sand volcanoes. *Journal of Geological Society.* November.

Ng, R. and J. Han. 1994. Efficient and effective clustering methods for spatial data mining. *Proceedings of the 20th International Conference on Very Large Databases.* 144–155, Santiago, Chile.

Niangoran-Bouah, N. 1984. *L'univers Akan des Poids a Peser l'Or* (The Akan World of Gold Weights). Abidjan, Côte d'Ivoire: Les Nouvelles Editions Africaines.

Nussbaumer, H. J. 1981. *Fast Fourier Transform and Convolution Algorithms.* New York, NY: Springer-Verlag.

Oberhettinger, F. 1973. *Fourier Transforms of Distributions and Their Inverses: A Collection of Tables.* New York. NY: Academic Press.

Onyefulu, I. 2007. *A Triangle for Adaora: An African Book of Shapes*. London, UK: Frances Lincoln Publishers.

Ottevanger, W., J. van den Akker and L. De Feiter. 2007. *Developing Science, Mathematics, and ICT Education in Sub-Saharan Africa: Patterns and Promising Practices*. Washington, DC: The World Bank.

Palacio, M., D. Sol and J. Gonzalez, 2003. Graph-based knowledge representation for GIS data. *Proceedings of the Fourth Mexican International Conference on Computer Science* (ENC '03).

Palter, R. 1993. *Black Athena*, Afro-centrism, and the history of science. *History of Science* 31, 93, part 3:227–287.

PAMOSTAR. 2007. Report of the 6th symposium on training and research on Pan African Mathematics Olympiads. Ouagadougou, Burkina Faso, November 5–10.

Pantel, P. and D. Lin. 2002. Discovering word senses from text. In *Proceedings of SIGKDD-01.* San Francisco, CA.

Papoulis, A. 1962. *The Fourier Integral and Its Applications.* New York, NY: McGraw-Hill.

Parrinder, G., ed. 1985. *World Religions: From Ancient History to the Present.* New York, NY: Facts on File Publications.

Peitgen, H-O, S. Dietmar, H. Jurgens and L. Yunker. 1992. *Chaos and Fractals: New Frontiers of Science.* Berlin: Springer-Verlag.

Peitgen, H-O and D. Saupe. 1988. *The Science of Fractal Images*. Berlin: Springer-Verlag.

Peitgen, H-O and P. H. Richter. 1986. *The Beauty of Fractals*. Berlin: Springer-Verlag.

Petito, A. L. 1982. Practical arithmetic and transfer: A study among West African tribesmen. *Journal of Cross-Cultural Psychology* 13:15–28.

Polyanin, A. D. and A. V. Manzhirov. 1998. *Handbook of Integral Equations*. Boca Raton, FL: CRC Press.

Postle, D. 1980. *Catastrophe Theory—Predict and Avoid Personal Disasters*. London, UK: Fontana Paperbacks.

Poston, T. and I. Stewart. 1998. *Catastrophe Theory and Its Applications*. New York, NY: Dover.

Powell, A. 2002. Ethnomathematics and the challenges of racism in mathematics education. *Proceedings of the Third International Mathematics Education and Society Conference*. Helsigør, Denmark, April 2–7, 2002.

Price, G. B. 1984. *Multivariable Analysis*. New York, NY: Springer-Verlag.

Puu, T. 2000. *Attractions, Bifurcations, and Chaos: Nonlinear Phenomena in Economics*, New York, NY: Springer.

Ramanathan, J. 1998. *Methods of Applied Fourier Analysis*. Boston, MA: Birkhäuser.

Ramirez, R. W. 1985. *The FFT: Fundamentals and Concepts*. Englewood Cliffs, NJ: Prentice-Hall.

Reagan, T. G. and I. Ntshoe. 1987. Language policy and Black education in South Africa. *Journal of Research and Development in Education* 20, 2:1–8.

Rees, C. S. 1981. Egyptian fractions. *Mathematical Chronicle* (now *New Zealand Journal of Mathematics*) 10, 1–2:13–30.

Reich, D. E. and D. B. Goldtsein. 1998. Genetic evidence for a Paleolithic human population expansion in Africa. *Proceedings of the National Academy of Sciences* 95:8119–8123.

Riordan, J. 1968. *Combinatorial Identities*. New York, NY: John Wiley and Sons, Inc.

Riordan, J. 1958. *An Introduction to Combinatorial Analysis*. New York, NY: John Wiley and Sons, Inc.

Ritter, J. November 1989. Prime numbers. *UNESCO Courier* 12–17.

Robins, G. and C. Shute. 1987. *The Rhind Mathematical Papyrus: An Ancient Egyptian Text*. London, UK: British Museum Publications, Ltd.

Robins, G. and C. Shute. 1985. Mathematical bases of ancient Egyptian architecture and graphic art. *Historia Mathematica* 12, 2:107–122.

Ruskai, M. B. et al., eds. 1992. *Wavelets and Their Applications*. Boston, MA: Jones and Bartlett.

Sallah, M. A. 1982. An investigation into the effectiveness of teaching mathematics in primary one of Gambian schools. African Studies in Curriculum Development and Evaluation No. 51. Nairobi, Kenya: Kenya Institute of Education, African Curriculum Organization.

Sander, J. M. Ester and H. Kriegel. 1998. Density-based clustering in spatial databases: A new algorithm and its applications. *Data Mining and Knowledge Discovery* 2, 2:169–194.

Sanns, W. 2000. *Catastrophe Theory with Mathematica: A Geometric Approach.* Osnabrück, Germany: Der Andere Verlag.

Sarhangi, R. n.d. Tessellation and tessellation exploration. Retrieved on July 11, 2008. Available at http://ages.towson.edu/gsarhang/tessellation%20Exploration.doc

Saunders, P. T. 1980. *An Introduction to Catastrophe Theory.* Cambridge, UK: Cambridge University Press.

Schey, H. M. 2005. *Div, Grad, Curl and All That: An Informal Text on Vector Calculus.* New York, NY: W. W. Norton and Company.

Schildkrout, E. and C. Keim, eds. 1990. *African Reflections: Art from Northeastern Zaire.* New York, NY: American Museum of Natural History.

Schlomiuk, D., ed. 1993. *Bifurcations and Periodic Orbits of Vector Fields* (NATO Advanced Study Institute and Seminare de mathematique supérieures on Bifurcations and Periodic Orbits of Vector Fields, Montreal, Canada, 1992). Boston, MA: Kluwer Academic Publishers.

Schmidl, M. 1915. Zahl und zahlen in Afrika. *Mitteilugen der Anthropologischen Gesellschaftin Wein* 45:165–209.

Schweinfurth, G. 1874. *The Heart of Africa: Three Years' Travels and Adventures in the Unexplored Regions of Central Africa from 1868 to 1871*, 2 vols. New York, NY: Harpers and Brothers.

Sears, J. W. April 2006. Earth tessellation II: Self-organized breakup of Gondwana. Retrieved on July 11, 2008. Available at http://www.mantleplumes.org/EarthTess2.html

Sears, J. W., G. M. St. George and J. C. Winne. March 2005. Continental rift systems and anorogenic magnetism. *Lithos* 80:147–154.

Seeman, E. C. 1977. *Catastrophe Theory—Selected Papers 1972–1977.* Reading, MA: Addison-Wesley.

Seidenberg, A. 1988. On the volume of a sphere. *Archive for History of Exact Sciences* 39, 2:97–119.

Seidenberg, A. and J. Casey. 1980/1981. The ritual origin of the balance. *Archive for History of Exact Sciences* 23, 3:179–226.

Sertima, I. V., ed. 1983. *Blacks in Science: Ancient and Modern.* New Brunswick, NJ: Transaction Books.

Setati, M. 2005. Teaching mathematics in a primary multilingual classroom. *Journal for Research in Mathematics Education* 36,5:447–466.

Setati, M. and J. Adler. 2001. Between languages and discourses: Language practices in primary multilingual mathematics classrooms in South Africa. *Educational Studies in Mathematics* 43, 3:243–269.

Shekhar, S., C. Lu, P. Zhang, and R. Liu. 2002. Data mining and selective visualization of large spatial datasets. *Proceedings of the 14th IEEE International Conference on Tools with Artificial Intelligence* (ICTAI '02).

Sherratt, J. A. and G. J. Lord. 2007. Nonlinear dynamics and pattern bifurcations in a model for vegetation stripes in semi-arid environments. *Theoretical Population Biology* 71:1–11.

Slomson, A. 1991. *An Introduction to Combinatorics.* London, UK: Chapman and Hall Limited.

Smelyakov, S. V., G. Stray and J. Wicherink. 2006a. The Auric Time Scale. *Astrotheos*. Retrieved on August 9, 2008. Available at http://www.ASTROTHEOS.narod.ru

Smelyakov, S. V., G. Stray and J. Wicherink. 2006b. The last multi-turns of the spiral of time before it rolls up to appear in new reality. *Astrotheos*. Retrieved on August 9, 2008. Available at http://www.ASTROTHEOS.narod.ru

Sneddon, I. N. 1995. *Fourier Transforms*. New York. NY: Dover.

Sogge, C. D. 1993. *Fourier Integrals in Classical Analysis*. New York, NY: Cambridge University Press.

Spiegel, M. R. 1974. *Theory and Problems of Fourier Analysis with Applications to Boundary Value Problems*. New York, NY: McGraw-Hill.

Stanley, R. P. 1997 & 1999. *Enumerative Combinatorics* vols. 1 & 2. Cambridge, UK: Cambridge University Press.

Steen, L. A., ed. 1988. *Introduction to Contemporary Mathematics*. New York: W. H. Freeman and Co.

Stein, S. K. and A. Szabó. 1994. *Algebra and Tiling: Homomorphisms in the Service of Geometry*. Washington, DC: The Mathematical Association of America.

Stein, E. M. and Weiss, G. L. 1971. *Introduction to Fourier Analysis on Euclidean Spaces*. Princeton, NJ: Princeton University Press.

Stewart, I. and M. Golubitsky. 1992. *Fearful Symmetry: Is God a Geometer?* Oxford: Blackwell.

Strogatz, S. 2000. *Non-linear Dynamics and Chaos: With Applications to Physics, Biology, Chemistry and Engineering*. London, UK: Perseus Books.

Swetz, F. J. 1984. Seeking relevance? Try the history of mathematics. *Mathematics Teacher* 77:54–62.

Tai, C-T. 1997. *General Vector and Dyadic Analysis: Applied Mathematics in Field Theory*. Washington, DC: IEEE Computer Society Press.

Tai, C-T. 1995. *A Historical Study of Vector Analysis*. Technical Report RL 915. Ann Arbor, MI: University of Michigan. Available at http://deepblue.lib/umich.edu/handle/2027.42/7868

The Internet Encyclopedia of Science. Lebombo bone. Retrieved on July 25, 2007 from http://www.daviddarling.info/encyclopedia/L/Lebombo_bone.html

Thom, R. 1989. *Structural Stability and Morphogenesis: An Outline of a General Theory of Models*. Reading, MA: Addison-Wesley.

Thompson, J. and T. Michael. 1982. *Instabilities and Catastrophes in Science and Engineering*. New York, NY: Wiley and Sons.

Tolstov, G. P. 1976. *Fourier Series*. New York, NY: Dover.

Trotter, W. T. 1992. *Combinatorics and Partially Ordered Sets: Dimension Theory*. Baltimore, MD: The Johns Hopkins University Press.

Tucker, A. 1984/1995. *Applied Combinatorics 2nd and 3rd eds*. New York, NY: John Wiley and Sons, Inc.

Turner, L. D. 1949/1969. *Africanisms in the Gullah Dialect*. New York, NY: Arno Press.

van der Waerden, B. L. 1983. *Geometer and Algebra in Ancient Civilizations*. Berlin, Germany: Springer.

van Heerden, J., C. Getz and H. Smuts (H. Smuts, ed.). 2005. *Africa Meets Africa: Making a Living through the Mathematics of Zulu Design*. Johannesburg, South Africa: David Krut Publishing.

van Lint, J. H. and R. M. Wilson. 2001. *A Course in Combinatorics 2nd ed.* Cambridge, UK: Cambridge University Press.

Vital, R and O. Skovsmose. 1997. *The End on Innocence: A Critique of Ethnomathematics*. Copenhagen, Denmark: The Royal Danish School of Educational Studies.

Waldrop, M. M. 1992. *Complexity: The Emerging Science at the Edge of Order and Chaos*. New York, NY: Simon and Schuster.

Walker, J. S. 1996. *Fast Fourier Transforms, 2nd ed.* Boca Raton, FL: CRC Press.

Washburn, D. 1990. *Style, Classification and Ethnicity: Design Categories on Bakuba Raffia Cloth*. Philadelphia, PA: American Philosophical Society.

Watkins, C. D. and L Sharp (M. Finlay, ed.). 1992. *Programming in 3 Dimensions*. San Mateo, CA: M&T Books.

Wegman, E. 2003. Visual data mining. *Statistics in Medicine* 22:1383–1397 plus 10 color plates.

Wegman, E. 1997. A Guide to Statistical Software. Available at http://www.galaxy.gmu.edu/papers/astr1.html

Weisstein, E. W. n.d. Feigenbaum Constant. Retrieved on August 08, 2008. Available at *MathWorld*—A Wolfram Web Resource: http://mathworld.wolfram.com/FeigenbaumConstant.html

Wells, R. A. 1997. Egypt. In C. F. walker, ed. *Astronomy Before the Telescope*. London, UK: British Museum Press.

Wells, R. A. 1994. Re and the calendars. In A. J. Spalinger, ed. *Revolutions in Time: Studies in Ancient Egyptian Calendrics*. San Antonio, TX: Van Siclen Press.

Wernham, B. n.d. Omweso: The royal Mancala game of Uganda—a general overview of current research. Retrieved on July 29, 2008. Available at http://www.omweso.org

Wickerhauser, M. V. 1994. *Adapted Wavelets Analysis from Theory to Software*. Wellesley, MA: A. K. Peters.

Wiens, E. G. n.d. Bifurcations and two dimensional flows. Egwald Web Services Ltd. Retrieved on August 7, 2008. Available at http://www.egwald.ca/nonlineardynamics/bifurcations.php

Wieting, T. W. 1982. *The Mathematical Theory of Chromatic Plane Ornaments*. New York, NY: Marcel Dekker, Inc.

Wilcox, A. R. 1984. *The Rock Art of Africa*. New York, NY: Holmes and Meier.

Williams, A. 1978. Change in mathematics education since the late 1950's—Ideas and realisation; Sierra Leone. *Educational Studies in Mathematics*. 9, 3:297–302.

Williamson, S. G. 1985. *Combinatorics for Computer Science*. Rockville, MD: Computer Science Press.

Wilson, III, E. J. and K. Wong. 2003. African information revolution: A balance sheet. *Telecommunications Policy* 27:155–177.

Woodcock, A., R. Edward and M. Davis. 1978. *Catastrophe Theory*. New York, NY: E. P. Dutton.

Wong, P., P. Whitney and J. Thomas. 1999. Visualizing association rules for text mining. In G. Wills and D, Keim, eds. *Proceedings of IEEE Information Visualization '99*. Los Alamitos, CA: IEEE CS Press.

Wong, R. and W. Shui. 2001. Utilizing multiple bioinformatics information sources: An XML database approach. *Proceedings of the Second IEEE International Symposium on Bioinformatics and Bioengineering.*

Woodson, C. G. 1933. *Mis-education of the Negro*. Washington, DC: The Associated Publishers, Inc.

Yang, R., X. Deng, M. Kafatos, C. Wang and X. Wang. 2001. An XML-based distributed metadata server (DIMES) supporting earth science metadata. *Proceedings of the 13th International Conference on Scientific and Statistical Database Management* 251–256.

Yoshida, K. 1968. *Functional Analysis*. New York, NY: Springer-Verlag.

Young, E. C. 1993: *Vector and Tensor Analysis 2nd ed.* New York, NY: Marcel Dekker, Inc.

Zaaiman, H. H. van der Flier and G. D. Thijs. 2000. Selection as contract to teach at the student's level: Experience from a South African mathematics and science foundation year. *Higher Education* 40, 1:1–21.

Zaslavsky, C. 1973/1999. *Africa Counts: Number and Pattern in African Culture*. Boston, MA: Prindle, Weber and Schmidt, Inc.

Zaslavaky, C. 1970. Black African traditional mathematics. *Math Teacher* 63, 4:345–356.

Zentella, A. C. 1997. *Growing up Bilingual: Puerto Rican Children in New York*. Malden, MA: Blackwell.

Zhusubaliyev, Z. T. and E. Mosekilde. 2003. *Bifurcations and Chaos in Piecewise-smooth Dynamical Systems*. River Edge, NJ: World Scientific.

WEB SITES

http://garsia.math.yorku.ca/ghana03/
http://garsia.math.yorku.ca/ghana04/
http://imagiware.com/mancala/
http://math.truman.edu/~thammond/history/SubSaharanAfrica.html
http://members.aol.com/afsci/kani.htm
http://tqd.advanced.org/3288/chaos.html
http://www.cohums.ohio-state.edu/comp/eglash.dir/afractal.htm
http://www.cohums.ohio-state.edu/comp/isgem.html
http://www.dd.chalmers.se/~gu94joli/icons.html
http://www.egypt-topics.com/Egypt-fashion-People/facts-about-egyptian-god-set.html
http://www.mat.uniroma1.it/~procesi/Kampala.html
http://www.math.buffalo.edu/mad/Ancient-Africa
http://www.ncsa.uiuc.edu/Edu/Fractal/Fgeom.html

http://www.osu.edu/org/awognet/
http://www.papres.com/virtual/index.html
http://www.southafrica.info/travel/cultural/border-cave.htm
http://www.students.uiuc.edu/~ag-ho/chaos/chaos.html
http://www.ubuntu-forum.com/showthread.php?t=836271

CPSIA information can be obtained
at www.ICGtesting.com
Printed in the USA
LVHW050201020319
609288LV00035B/1010

9 780761 85348C